Women in Engineering and Science

The Springer Women in Engineering and Science series highlights women's accomplishments in these critical fields. The foundational volume in the series provides a broad overview of women's multi-faceted contributions to engineering over the last century. Each subsequent volume is dedicated to illuminating women's research and achievements in key, targeted areas of contemporary engineering and science endeavors. The goal for the series is to raise awareness of the pivotal work women are undertaking in areas of keen importance to our global community.

Amanda Weissman
Editor

Women in Advanced Computing and Software for Defense

Springer

Editor
Amanda Weissman
Blackwood, NJ, USA

ISSN 2509-6427 ISSN 2509-6435 (electronic)
Women in Engineering and Science
ISBN 978-3-032-17465-9 ISBN 978-3-032-17466-6 (eBook)
https://doi.org/10.1007/978-3-032-17466-6

This Springer imprint is published by the registered company Springer Nature Switzerland AG
The registered company address is: Gewerbestrasse 11, 6330 Cham, Switzerland

Foreword

The strategic environment isn't getting simpler. Adversaries are leveraging technological acceleration, often with fewer bureaucratic constraints and faster decision cycles. From Ukraine, where commercial drones are modified for battlefield use, to Israeli military operations, where AI-powered systems process intelligence at speeds orders of magnitude faster than human capacity—innovation cycles are measured in days and weeks rather than months and years [1–4]. The question isn't whether to accelerate our adoption of advanced computing and software capabilities—it's whether we can do so in ways that are secure, trustworthy, and aligned with our values.

This book addresses that central challenge across critical elements of Advanced Computing and Software. As one of 14 critical technology areas essential to national security, fielding software-enabled platforms and weapons systems will define capability across all domains [5]. Yet despite clear need to accelerate fielding, systems are not advancing at the required speed. Over the past decade, Government Accountability Office (GAO) assessments have shown we're actually moving backward—costs have increased and delivery has slowed [6, 7]. It is incontrovertible that we need fundamental transformation in how we develop, acquire, deploy, and sustain software-intensive systems, grounded in the pragmatic reality of federal procurement and the critical necessity of balancing safety with speed and agility. The question is how. Can we build systems that warfighters will use, program managers can acquire, and maintainers can sustain? Can we create organizational cultures that value speed without sacrificing rigor, embrace experimentation while managing risk, and leverage commercial innovation while protecting national security?

This book goes through critical elements of that transformation from data to artificial intelligence (AI) to modeling and simulations—balancing theoretical frameworks with insights gleaned from the authors' experiences. While each individual chapter in this book provides critical insights, what makes this volume particularly valuable is the implementation guidance from people who have done this work—authors from large defense contractors, government service providers, and Federally Funded Research and Development Centers, united by their commitment to solving problems at the intersection of policy and practice, strategy and execution, innovation and security. These experts—all of whom happen to be women—represent a strategic advantage in and of itself. Beginning this book by understanding contributions from Admiral Grace Hopper to Margaret Hamilton highlights how critical it is to not only to honor their work but to motivate the next generation. The chapter Legacy of Women in Defense Computing highlights how women have been instrumental architects of defense computing since its inception. This book continues that legacy, showcasing modern leaders actively shaping defense innovation.

Research consistently demonstrates that diverse teams outperform homogeneous ones in innovation, problem-solving, and identifying failure modes [8]. In defense contexts involving complex socio-technical systems, adversarial thinking, and rapidly evolving threats, this isn't just about better outcomes—it's about mission success. With women comprising just over a quarter of the US workforce in computer and mathematical occupations, and even lower representation in defense sectors, we risk missing critical talent [9]. Any system that fails to include the full range of available perspectives will make worse decisions and that will affect national security and human lives. This isn't an issue about fairness (though that matters as well); this is a capability issue that is critical to ensuring warfighters have the benefit of the best talent available to support their mission.

Building on that foundational talent requirement, the chapter Speed of Need: Unifying Agile, DevSecOps, MBSE, MOSA, Digital Engineering and AI for Defense Systems highlights the criticality of integration across the entire digital engineering value stream. By synthesizing Agile, DevSecOps, Model-Based Systems Engineering, and generative AI, it is possible to dramatically improve software security while maintaining deployment velocity [10]. The challenge isn't knowing about these practices—it's executing them within constraints of the defense acquisition process, security classification requirements, and legacy infrastructure. Recent research from the Software Engineering Institute and US Space Force case studies show that when agile practices are properly adapted to defense contexts, they significantly reduce development cycles and improve responsiveness [11, 12]. Successful adoption requires understanding both the technical and organizational dimensions of change.

Rapidly delivering that capability in the era of AI transformation depends on data; thus, data strategy plays a critical role. The 2020 DoD Data Strategy recognized data as foundational to AI, machine learning, and analytics; in 2021 the Department of Defense decreed data a strategic asset [13, 14]. The chapter Harnessing Data for Defense: Challenges and Solutions provides a roadmap for building infrastructure that is interoperable across legacy systems, flexible enough to enable AI models, secure enough to protect classified information, and compliant across multiple frameworks—addressing governance structures, cataloging systems, and access controls that work in classified environments.

Maximizing the use of that data happens through the use of advanced AI applications that must be deployed at scale. This requires more than algorithmic sophistication. The chapter The Artificial Intelligence Factory: Concept, Implementation, and Impacts provides a framework for the organizational culture changes needed to accelerate technology delivery—empowering domain experts and building trust through human-centered design. In military applications where AI systems operate in contested environments with life-or-death consequences, ensuring accountability and trustworthiness under adversarial pressure is critical. The chapter describes practical implementation pathways, tool selections, and governance structures for responsible AI deployment.

Modeling and simulation in classified environments presents unique challenges. The chapter Navigating Computational Challenges and Security Considerations in Defense Industry Modeling and Simulation provides invaluable insights on developing M&S frameworks when you can't leverage the cloud services, open-source libraries, or collaborative tools available in unclassified contexts.

In short this book focuses on the defining reality of twenty-first-century warfare: technology that can deliver decisive effects at the tactical edge is evolving at an accelerating rate. Military advantage increasingly lies with whoever can deploy the best technology fastest. The challenge defense organizations must solve: delivering advanced computing and software capabilities at the speed of operational need. The technical, organizational, cultural, and representational transformations described in these chapters are essential to delivering capabilities at the speed of need. And so, I hope this volume inspires action. The mission is real, the challenges are hard, and the work matters. Ensuring innovation—and the voices that drive that innovation—is not a luxury but a national security imperative.

Dr. Radha Iyengar Plumb
Senior Fellow
Wharton Accountable Artificial
Intelligence Lab
Distinguished Visiting Fellow
Perry World House
University of Pennsylvania
Former Chief Digital and Artificial
Intelligence Officer
US Department of Defense
Washington DC
United States of America

References

1. Atlantic Council, "Drone Superpower: Ukrainian Wartime Innovation Offers Lessons for NATO," May 2025, https://www.atlanticcouncil.org/blogs/ukrainealert/drone-superpower-ukrainian-wartime-innovation-offers-lessons-for-nato/.
2. Chatham House, "What Ukraine Can Teach Europe and the World About Innovation in Modern Warfare," March 2025, https://www.chathamhouse.org/2025/03/what-ukraine-can-teach-europe-and-world-about-innovation-modern-warfare.
3. Belfer Center for Science and International Affairs, "Advancing in Adversity: Ukraine's Battlefield Technologies and Lessons for the U.S.," July 2023, https://www.belfercenter.org/publication/advancing-adversity-ukraines-battlefield-technologies-and-lessons-us.

4. NPR, "Israel Is Using an AI System to Find Targets in Gaza. Experts Say It's Just the Start," December 2023, https://www.npr.org/2023/12/14/1218643254/israel-is-using-an-ai-system-to-find-targets-in-gaza-experts-say-its-just-the-st.
5. Department of Defense, "Critical Technology Areas," accessed October 2025, https://www.cto.mil/osc/critical-technologies/.
6. U.S. Government Accountability Office, "Weapon Systems Annual Assessment: Knowledge Gaps Pose Risks to Sustaining Recent Positive Trends," GAO-18-360SP, April 2018, https://www.gao.gov/products/gao-18-360sp.
7. U.S. Government Accountability Office, "Weapon Systems Annual Assessment: Programs Are Not Consistently Implementing Practices That Can Help Accelerate Acquisitions," GAO-23-106059, June 2023, https://www.gao.gov/products/gao-23-106059.
8. Lorenzo, Rocío, Nicole Voigt, Miki Tsusaka, Matt Krentz, and Katie Abouzahr, "How Diverse Leadership Teams Boost Innovation," Boston Consulting Group, January 2018; Rock, David, and Heidi Grant, "Why Diverse Teams Are Smarter," Harvard Business Review, November 2016.
9. Society of Women Engineers, "Employment Statistics," 2024, https://swe.org/research/2024/employment/.
10. Mohan, Vivek, and Lotfi ben Othmane, "Challenges and Solutions when Adopting DevSecOps: A Systematic Review," Information and Software Technology 141 (2022): 106741; Software Engineering Institute, Carnegie Mellon University, "Continuous Deployment of Capability," accessed October 2025, https://www.sei.cmu.edu/our-work/devsecops/.
11. Gracias, Maryam H., and Stephanie C. Kestner, "Agile vs. Waterfall in Aerospace and Defense," ITEA Journal 45, no. 4 (2024); https://itea.org/journals/volume-45-4/agile-and-waterfall-perceptions/.
12. University of Southern California Viterbi School of Engineering, "Mission Critical: Improving Software Development for Space Defense," December 2024, https://viterbischool.usc.edu/news/2024/11/mission-critical-improving-software-development-for-space-defense/.
13. Department of Defense, "DoD Data Strategy," 2020.
14. Deputy Secretary of Defense Kathleen Hicks, "Creating Data Advantage," memorandum, May 5, 2021, https://media.defense.gov/2021/May/10/2002638551/-1/-1/0/DEPUTY-SECRETARY-OF-DEFENSE-MEMORANDUM.PDF.

Acknowledgements

This work would not have been possible without the many individuals who supported and encouraged us throughout this journey. We are indebted to those whose guidance, feedback, and belief in our vision enabled us to bring this book to completion.

I wish to extend particular recognition to Karen Roth, whose insight and initiative were the catalyst for this effort. Without her, the very idea for this volume would not have existed.

To the chapter authors, thank you for daring to imagine what we could accomplish together. Your expertise, dedication, and willingness to collaborate transformed this project from concept to reality.

To Adam Weissman, I am especially grateful for your steady support, thoughtful technical discussions, and recommendations, all of which helped strengthen this work at every stage.

Finally, I acknowledge with gratitude the broader community of colleagues, mentors, friends, and family whose encouragement and example built us up, providing the foundation upon which we could strive for—and ultimately achieve—greatness.

Contents

Introduction

Amanda Weissman

Abstract This introduction explores the critical role of Advanced Computing and Software in the U.S. Department of Defense's strategy to maintain national security and military superiority. The chapter explains the role of the following chapters within the broader context of evolving warfare, where digital, cyber, and autonomous systems reshape the battlefield and demand rapid innovation. It outlines the Department of Defense's identification of 14 critical technology areas, with a focus on Advanced Computing and Software, including big data, artificial intelligence, modeling and simulation, and secure software delivery. Ultimately, this chapter frames the foundation for the book: showcasing women leaders who are shaping the future of defense through innovation in advanced computing and software.

1 Introduction

When most people think of the United States of America's defense industry, they don't immediately picture women engineers and computer scientists at the forefront of technical innovation. The field is often associated with "old boys' clubs" with many of its members having grown up as boys playing military games. This book will show, however, women have been an integral part of the history of the defense industry, especially in advanced computing and software. From trailblazers, such as Rear Admiral Grace Murray Hopper, to unsung pioneers like Judy Clapp, women have contributed as inventors, visionaries, and changemakers. Today, their legacy continues through the authors of these chapters: modern leaders driving innovation, securing the future, and transforming the defense industry.

In this book, we use the term women in defense to refer to women in engineering and science within the U.S. defense industrial base as well as government employees in defense-related roles. The U.S. defense industrial base includes large prime defense contractors, government services and solutions providers, and Federally Funded Research and Development Centers (FFRDCs), to name a few. We

A. Weissman (✉)
Moorestown, NJ, USA
e-mail: amanda.j.weissman@gmail.com

A. Weissman (ed.), *Women in Advanced Computing and Software for Defense*, Women in Engineering and Science, https://doi.org/10.1007/978-3-032-17466-6_1

also acknowledge the broader ecosystem of women contributing to national defense, including those in academia conducting defense-funded research, entrepreneurs and small businesses supported by programs such as small business innovation research (SBIR) and small business technology transfer (STTR).

Research quantifying the number of women in defense is limited, but trends can be inferred from broader data on women in Science, Technology, Engineering, and Math (STEM). As of 2023, women make up just 26.9% of the workforce in all computer and mathematical occupations in the United States [1]. Representation of women is even lower within the defense sector. Women comprise only 23–25% of the workforce at the top five U.S. defense contractors—a figure that includes all job functions, not just STEM roles [2]. In 2019, the Annual Report on the Federal Workforce published a special topic on Women in STEM. This study found women accounted for 29.3% of STEM federal employees [3], not limited to the defense industry. More specifically in defense, in 2020 at the Air Force Research Laboratory (AFRL), women made up just 17.56% of the workforce [4]. From this data, it's reasonable to assume women in defense make up between 17 and 29% of the workforce, depending on other factors such as specific occupations and sector (federal or industry). The AFRL study was one example also showing women in defense-related roles experience higher attrition rates than their male counterparts [4]. These statistics underscore the importance of highlighting the exceptional women who are not only contributing to, but also leading, innovation in our national defense ecosystem.

What all these women bring to the defense industry is their passion for the mission and their exceptional technical talents. The national security of the United States depends on continuous technological innovation—an area where women make significant and ongoing contributions. This book focuses specifically on advanced computing and software, one of the 14 critical technology areas identified by the Department of Defense's Office of Research and Engineering as essential to maintaining national security and military superiority while driving innovation. This critical technology area encompasses advanced computing capabilities, modern software systems, and the technologies that support and enable them.

2 The Strategic Landscape: National Security and the Digital Battlefield

There is a consistent theme across the multitude of reports, articles, and assessments looking at national security: traditional warfare has evolved and now needs to encompass the digital battlefield and we must accelerate innovation to keep pace with the ever-evolving threat landscape.

2.1 *Evolving Nature of Warfare: From Kinetic to Digital and Cyber*

The hardware of the past World Wars, aircraft, aircraft carriers, tanks, and submarines, may not have changed significantly but the tactics on how we use those assets has evolved and will continue to do so. Where traditional warfare focused on firepower, today, the success of a conflict increasingly relies on the availability, accuracy, and ability to use information [5]. This data will not exist in isolated silos; rather, it is essential that it be interoperable across the entire battlefield. The critical role of data and information on the battle field can be visualized through the Data-Information-Knowledge-Wisdom (DIKW) pyramid in Fig. 1, which highlights how raw data is transformed into actionable insight [6]. The evolution of military information systems illustrates this need: beginning with Command and Control (C2), expanding to Command, Control, and Communications (C3), then to Command, Control, Communications, and Computers (C4), and now encompassing Command, Control, Computers, Communications, Cyber-Defense, Combat Systems, Intelligence, Surveillance, and Reconnaissance (C6ISR). The next evolution of C6ISR is the Joint All Domain Command and Control (JADC2), which enables information to be delivered at the "speed of relevance" while coordinating across all domains, land, sea, air, space, and cyber [7].

This transition toward faster and larger collections of information has been shown by the shift in procurement from expensive and large radars of yesterday to the inexpensive, expendable drones of today to collect situational awareness data. The evolution of technology has also changed the modern battlefield by making it more accessible to smaller countries and non-state actors. Low-cost technology with a lower barrier to entry has also enabled low-cost cyber warfare [5]. Additionally, the

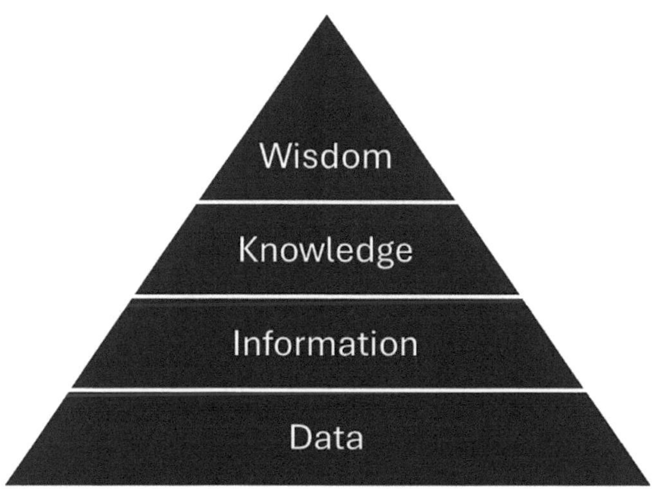

Fig. 1 Data-Information-Knowledge-Wisdom (DIKW) pyramid [6]

effectiveness of our traditional weapons also decreases as today's equipment are capable of spoofing or imitating a trusted entity [5]. This combination of improved sensors, automation, and artificial intelligence (AI) with hypersonic weapons and other advanced technologies will produce more accurate, better connected, faster, longer range, and more destructive weapons, primarily available to the most advanced militaries but some within reach of smaller state and non-state actors. The proliferation and diffusion of these systems over time will make more assets vulnerable, heighten the risk of escalation, and make combat potentially more deadly, though not necessarily more decisive [5].

Furthermore, autonomous systems and AI play an ever-increasing role in warfare. While not new weapons themselves, autonomous systems are enabling technologies that allow platforms to operate with decreasing levels of human interaction, for extended periods, and in increasingly hazardous environments [8]. Examples include uncrewed vehicles operating in air, land, sea, or undersea that extend the battlefield further through swarming behaviors and as lethal autonomous weapons.

Cyber-attacks are also quickly evolving the nature of warfare. Cyber-attacks have a long history in defense beginning with the first computer virus, Creeper, detected on ARPANET in 1971 [9]. More recently, the United States Cyber Command (USCYBERCOM), having existed since the 1990s, was elevated to a combatant command in 2017 to improve the U.S.'s ability to operate in cyberspace [10]. Meanwhile, the line between military and civilian cybersecurity continues to blur, cyberwarfare today includes eroding citizens' trust in their own government and communities, eroding morale, and disrupting day-to-day life even prior to hostilities. For example, even prior to hostilities Russian defaced Ukrainian government webpages with false messages and disrupted the Ukrainian power grid [11]. Cyber-attacks can also impact the physical world too, such as with the Stuxnet virus, which damaged centrifuges in an Iranian nuclear facility [10].

According to Healey's Framework of Offensive Cyber Operations in Wartime, there are five types of cyber intent that happen before and during hostilities [11]:

- Exploiting information.
- Attacking information, networks, and IT systems.
- Attacking trust in institutions or erode morale.
- Attacking trust in military information or systems.
- Defeating physical infrastructure or weapon systems.

This framework illustrates that warfare may originate in cyberspace but can evolve into kinetic operations or a blend of cyber and kinetic operations [10]. From the first computer virus in 1971, to over six million cyber-attacks on the Pentagon in 2008 alone [10], the scope and scale of cyber threats have grown dramatically. With the rise of AI and "bots," this trend is likely accelerating.

As the battlefield evolves, engineers and scientists working in advanced computing and software consider cybersecurity throughout the lifecycle of a system because information is our most valuable asset. Defense begins with safeguarding design information during development and continues through operational deployment to ensure resilience against cyber-attacks.

3 Introduction to the 14 Critical Technology Areas

In 2023, the National Defense Science and Technology Strategy identified 14 critical technology areas vital to maintaining the United States' national security and these are grouped into three areas [12]:

1. **Seeds of Emerging Opportunity**
   (a) **Biotechnology**—Using living systems to produce technologies and capabilities.
   (b) **Quantum Science**—Study of physical properties at small atomic scales.
   (c) **Future Generation (FutureG) Wireless**—Wireless technologies enable military operations and free and open Internet.
   (d) **Advanced Materials**—Innovative new materials and novel manufacturing techniques.
2. **Effective Adoption Areas**
   (a) **Trusted Artificial Intelligence (AI) and Autonomy**—Applying software to perform tasks that currently require human intelligence.
   (b) **Integrated Network System-of-Systems**—Real-time dissemination of information across the Department of Defense enabling effective command, control, and communication.
   (c) **Microelectronics**—Circuits and components within electronic systems that need to be secured and produced.
   (d) **Space Technology**—Space flight, communications, and other technology to maintain space operations.
   (e) **Energy Resilience**—Reliable, flexible power for mission success.
   (f) **Advanced Computing and Software**—Ubiquitous throughout the Department of Defense enabling transformation in warfighting capabilities.
   (g) **Human–Machine Interfaces**—Technology related to human–machine teaming and augmented reality.
3. **Defense-Specific Areas**
   (a) **Directed Energy**—Lasers, high power microwaves, and high-energy particle beams used for warfare.
   (b) **Hypersonics**—Systems flying within atmosphere at five times the speed of sound.
   (c) **Integrated Sensing and Cyber**—Providing an advantage in highly contested environments.

This book focuses on advanced computing and software, with particular emphasis on big data, AI, modeling and simulation, and rapid capability delivery. These areas are not only central to modern defense innovation but also showcase the contributions of women engineers and scientists who are leading and shaping the development of these technologies. Highlighting their work underscores the impact of women in securing national defense, driving technology breakthroughs, and transforming the traditionally male-dominated defense industry. While this book focuses specifically on advanced computer and software, we would like to acknowledge the multitude of strategies and implementation plans that exist today within the sphere of defense, including:

- U.S. Department of Defense Software Modernization Implementation Plan (FY25–26).
- Fulcrum: DoD IT Advancement Strategy.
- DoD Zero Trust Strategy.
- DoD Data Analytics and AI Adoption Strategy.
- DoD Cyber Workforce Strategy and Implementation Plan.
- DoD Software Science and Technology Strategy and Implementation Plan.
- DoD Cloud FinOps Strategy.
- DoD OCONUS Cloud Strategy.
- DoD Records Strategy.
- Application Program Interface Technical Guidance.

3.1 Defining Advanced Computing and Software

In this book, we will dive deeper in Advanced Computing and Software, one of the 14 critical technology areas discussed previously. But what is advanced computing and software? According to the Department of Defense,

> Advanced computing and software technologies include supercomputing, cloud computing, data storage, computing architectures, and data processing. Software is ubiquitous throughout the Department, but the speed at which software develops outpaces the Department's ability to stay up to date. The Department must rapidly modernize its legacy software systems with resilient, affordable, and assured new software that has been designed, developed, and tested using processes that establish confidence in its performance. The Department must migrate to a Development-Security-Operations (DevSecOps) approach in its software development and evolve to a model of continuous development, continuous test, and continuous delivery. The Department must leverage modular open system architecture approaches to isolate hardware from software and enable rapid upgrades to secure processors [13].

Taking it a step further, according to the Research and Development Roadmap for Advanced Computing and Software, software is critical to almost every weapon system and is an enabler for other critical technologies such as AI and quantum computing. One goal of the Advanced Computing and Software critical technology area is to make a measurable impact on the time to field software to the warfighter. In the past, software updates have taken 5–8 years to reach the warfighter. In the future, the defense industry would like to reach parity with the commercial sector with updates reaching the user as needs arise, even multiple times a day [14].

To make this transition there are four sub-goals identified by the roadmap [14]:

1. **Speed-of-need software** that enables software to evolve quickly in the field. The speed-of-need is how quickly systems—or, in this case, software—must be updated to maintain an advantage over adversaries [15].

2. **Hybrid, adaptable computing architectures** that are more powerful use diverse computing approaches and enable flexibility.
3. **Workforce effectiveness at scale** which acknowledges that technology alone will not meet these goals but we also need the talent and enable them through AI to work efficiently.
4. The foundation of all these sub-goals is that it must be based on **secure data**.

While this book does not tackle all the areas put forth in Research and Development Roadmap for Advanced Computing and Software, it focuses on critical components that drive rapid innovation and operational impact: software factory ecosystems to accelerate software delivery, AI-enabled software generation, modeling and simulation, and secure data practices. By highlighting these areas, the book underscores how advanced computing and software transform defense capabilities and showcases the contributions of women engineers and scientists in driving this innovation. Let us define a few key themes used throughout the chapters of this book.

Software factory is "defined as a collection of people, tools, and processes that enables teams to continuously deliver value by deploying software to meet the needs of a specific community of end users" [16]. Today, software factories exist and are continuing to grow throughout the defense industry. Some are established directly by federal agencies, others through industry companies, and some are a collaborative partnership between the two.

Modern software development practices often refer to agile or iterative software development and DevSecOps. Popular software development approaches have sought to shorten feedback loops and accelerate delivery through frameworks such as Extreme Programming, Scrum, and the Scaled Agile Framework (SAFe). The intention of them all stems from the "Manifesto for Agile Software Development" with the goal to deliver working software faster to the end user. DevSecOps is the combination of software development (Dev), integrated cybersecurity (Sec), and operating software (Ops). DevSecOps enables the cross collaboration of people dedicated to these disciplines to enable the delivery of secure, high-quality software that meets the goals of all three disciplines [16].

Foundational to all of this is **cybersecurity**. The core concept of cybersecurity is to ensure confidentiality, integrity, and availability. This means that the data or information within these information systems is only accessed by those who need to know, that users can trust the integrity of the data, and that the data and systems are available when they need them. Cybersecurity enables the development and operation of secure software but touches many aspects of software including the supply chain, chip-based security, identity management, network security, endpoint security, threat detection and response, cloud security, data security, email security, application security, unified threat management, vulnerability management, managed security services, post-breach response services, and risk and compliance services [17]. Evolution of cybersecurity has led to a goal of "Zero Trust" within the DoD which means that no system or user will be implicitly trusted as exist in today's perimeter-based defenses but instead systems and users will always be verified [18]. Throughout these chapters, there is an underlying assumption of data protection, reflected in references to security clearances required for accessing protected information and the separation of computer systems into unclassified and classified environments.

Another area discussed in multiple chapters is **artificial intelligence (AI)**. The DoD defines AI as "the ability of machines to perform tasks that normally require human intelligence...whether digitally or as the smart software behind autonomous physical systems" [19]. Examples include performing complex tasks that typically require human intelligence such as pattern recognition, learning from experience, drawing conclusions, or making predictions [19]. For advanced computing and software, AI can refer to the tasks performed by engineers and scientists to develop defense systems as well as the defense systems themselves when they incorporate AI, such as the autonomous systems previously discussed.

The overview and themes presented here lay the foundation for understanding advanced computing and software in defense. Familiarity with the definitions, roadmaps, and broader context surrounding these technologies provides essential background for the chapters that follow, helping readers appreciate how the depth of research and innovation highlighted in this book advances the defense technology roadmap.

4 Innovation Ecosystem: Who Builds the Future?

Innovation in advanced computing and software is not driven by the federal government alone. Across the United States, large and small defense contractors, startups, academic institutions, and non-profit organizations collaborate to develop the technologies that shape the future. The U.S. Defense Industrial Base: Background and Issues for Congress demonstrates the vast network of organizations that can and do participate in the development, deployment, and maintenance of advanced computing and software to support defense in the U.S. [20]. It can also be seen that the Defense Industrial Base is not stagnant, and it continues to evolve and grow. Examples of this include the establishment and growth of software factories which "emerged organically as individuals and programs recognized the need to adopt commercial best practices to meet emerging defense priorities" [21]. In this book, you will hear from authors representing a range of backgrounds in the defense industrial base, including large prime defense contractors, government services and solutions providers, and Federally Funded Research and Development Centers (FFRDCs).

One important theme that we do not explore in depth in this book—but which deserves acknowledgment—is the challenge of acquisition. The Department of Defense has long faced difficulties shifting from a hardware-centric to a software-centric acquisition model. This transition has slowed the ability to update software and software-focused systems efficiently. To help address these challenges, the DoD established the Software Acquisitions Pathway, aimed at streamlining and modernizing software acquisition processes [22].

5 Future Trends and Challenges

Advanced computing and software is a continuously accelerating area of innovation. In 1965, the co-founder of Fairchild Semiconductor and Intel, Gordon Moore, noted that the number of components per integrated circuit had been doubling every year. In 1975, he revised his statement to a prediction looking to the next decade that it would continue to double every 2 years and has since been known as Moore's Law [23]. The implications of this have been that computing power and software accelerated every year. While today it is debated whether or not Moore's Law is still viable, the evolution of quantum computing, cloud computing, and AI mean that advanced computing and software will not only remain relevant but will still continue to accelerate. In modern warfare, adversaries are taking advantage of the continued pace of technology and national security interests must continue.

At the time of writing, the U.S. government continues to prioritize the advancement of computing and software through executive orders, Department of Defense initiatives, AI integration contracts, and other strategic efforts. This underscores both the critical importance of this field and the rapid pace at which it is evolving. While this chapter cannot capture the full complexity and ever-evolving nature of advanced computing and software and how it meets the needs of tomorrow's warfighter, we hope it has given you enough background to appreciate and be inspired by the critical work by the women in these upcoming chapters.

References

1. Society of Women Engineers (2025) U.S. Employment. Available at: https://swe.org/research/2025/us-employment/. Accessed 14 Sept 2025
2. Times Aerospace 'The growing role of Arab women in defence', *Times Aerospace*. Available at: https://www.timesaerospace.aero/news/defence/the-growing-role-of-arab-women-in-defence. Accessed 14 Sept 2025
3. U.S. Equal Employment Opportunity Commission (EEOC) (2022) *Special Topics Annual Report: Women in STEM—Supplement to the Annual Report on the Federal Workforce Fiscal Year 2019*. Washington, DC: EEOC. Available at: https://www.eeoc.gov/sites/default/files/2022-07/Women%20in%20STEM%20Special%20Topics%20Report%20FY%2020%20Final.pdf. Accessed 14 Sept 2025
4. Cook, G.M. and Wallentine, S.M. (2022) Women in Science, Technology, Engineering and Mathematics in the United States Air Force Research Laboratory: Factors Associated with Attrition and Promotion. Defense Technical Information Center. Available at: https://apps.dtic.mil/sti/trecms/pdf/AD1170348.pdf. Accessed: 14 Sept 2025
5. National Intelligence Council (2021) *The Future of the Battlefield*. NIC-2021–02493. Office of the Director of National Intelligence. Available at: https://www.dni.gov/files/images/globalTrends/GT2040/NIC-2021-02493--Future-of-the-Battlefield--Unsourced--14May21.pdf. Accessed 14 Sept 2025
6. Rowley, J. (2007) 'The wisdom hierarchy: representations of the DIKW hierarchy', *Journal of Information Science*, 33(2), pp. 163–180. https://doi.org/10.1177/0165551506070706. Accessed 4 Oct 2025

7. Beebe, E., 2024. 'C4ISR & JADC2: Navigating the Next Frontier in Military Command and Control'. *IDGA*. Available at: https://www.idga.org/command-and-control/articles/c4isr-jadc2-the-next-frontier-in-military-command-and-control. Accessed 14 Sept 2025
8. National Intelligence Council (2021) *The Future of the Battlefield*. NIC-2021–02493. p. 5. Office of the Director of National Intelligence. Available at: https://www.dni.gov/files/images/globalTrends/GT2040/NIC-2021-02493--Future-of-the-Battlefield--Unsourced--14May21.pdf. Accessed 14 Sept 2025
9. History of Information, 2025. *The Creeper Worm, the First Computer Virus*. Available at: https://www.historyofinformation.com/detail.php?entryid=2860. Accessed 21 Sept 2025
10. Plazas, C., 2021. Information Crossroads: Intersection of Military and Civilian Interpretations of Cyber Attack and Defense. *University of Cincinnati Intellectual Property and Computer Law Journal*, 5(1), Article 2. Available at: https://scholarship.law.uc.edu/cgi/viewcontent.cgi?article=1031&context=ipclj. Accessed 21 Sept 2025
11. Healey, J., 2024. Cyber Effects in Warfare: Categorizing the Where, What, and Why. *Texas National Security Review*, 7(4), pp.37–50. Available at: https://tnsr.org/2024/08/cyber-effects-in-warfare-categorizing-the-where-what-and-why/. Accessed 21 Sept 2025
12. Office of the Under Secretary of Defense for Research and Engineering (OUSW (R&E)). (n.d.). *OSC Critical Technology Areas*. Department of War. Available at: https://www.cto.mil/osc/critical-technologies/. Accessed 21 Sept 2025
13. Office of the Under Secretary of Defense for Research and Engineering (OUSD(R&E)). (n.d.). *DoD Critical Technology Area Roadmaps*. Available at: https://dod-critical-technology-area-roadmaps.zoiclabs.io/. Accessed 21 Sept 2025
14. Office of the Under Secretary of Defense for Research and Engineering (OUSD(R&E)). (2024). *Research and Development Roadmap for Advanced Computing and Software (ACS) – DOPSR APPROVED*. Department of Defense. Available at: https://dod-cta.s3.us-west-2.amazonaws.com/roadmaps/(U)%20ACS%20Distro%20A%20Roadmap%20-%20DOPSR%20APPROVED.pdf. Accessed 21 Sept 2025
15. Tipton, D. (2019). 'The speed of need'. *Peterson and Schriever Space Force Base*. Available at: https://www.petersonschriever.spaceforce.mil/Newsroom/Commentaries/Display/Article/1927072/the-speed-of-need/. Accessed 21 Sep. 2025
16. Department of Defense Chief Information Officer (DoD CIO). (2024). *DoD Enterprise DevSecOps Fundamentals v2.5*. Department of Defense. Available at: https://dodcio.defense.gov/Portals/0/Documents/Library/DoD%20Enterprise%20DevSecOps%20Fundamentals%20v2.5.pdf. Accessed 21 Sept 2025
17. Global Defence Technology. (2024). *The impact of cybersecurity on the defence sector*. Issue 148, August 2024. Available at: https://defence.nridigital.com/global_defence_technology_aug24/impact-cybersecurity-defence. Accessed 21 Sept 2025
18. Defense Acquisition University (DAU). (n.d.). *Zero Trust*. Available at: https://www.dau.edu/zerotrust. Accessed 21 Sept 2025
19. Defense Acquisition University (DAU). (n.d.). *DAI AI Toolkit: AI Descriptors and Definitions Reference Chart*. Available at: https://www.dau.edu/sites/default/files/Migrated/CopDocuments/DAI%20AI%20Toolkit%20-%20AI%20Descriptors%20and%20Definitions%20Reference%20Charts.pdf. Accessed 21 Sept 2025
20. Nicastro, L.A. (2024). *The U.S. Defense Industrial Base: Background and Issues for Congress* (CRS Report R47751). Congressional Research Service. Available at: https://www.congress.gov/crs-product/R47751. Accessed 21 Sept 2025
21. Department of Defense Chief Information Officer (DoD CIO). (2025). *The State of DevSecOps*. U.S. Department of Defense. Available at: https://dodcio.defense.gov/Portals/0/Documents/Library/DevSecOpsStateOf.pdf. Accessed 21 Sept 2025
22. Department of Defense. (2025). *Directing Modern Software Acquisition to Maximize Lethality*. Memorandum from the Secretary of Defense, 6 March 2025. Available at: https://media.defense.gov/2025/Mar/07/2003662943/-1/-1/1/DIRECTING-MODERN-SOFTWARE-ACQUISITION-TO-MAXIMIZE-LETHALITY.PDF. Accessed 21 Sept 2025

23. Arcuri, G. & Shivakumar, S., 2022. *Moore's Law and Its Practical Implications*. Center for Strategic and International Studies. Available at: https://www.csis.org/analysis/moores-law-and-its-practical-implications. Accessed 21 Sept 2025

Legacy of Women in Defense Computing

Tuyet-Hanh Nguyen Schnell

Abstract This chapter presents the history and impact of pioneering women in defense and computing. Ada Lovelace is recognized for her early vision of computer programming, while Admiral Grace Hopper advanced programming languages and influenced modern computing. Margaret Hamilton led the development of reliable software for the Apollo missions. Elizabeth "Jake" Feinler created early internet naming conventions that shaped today's digital infrastructure. The ENIAC Six—Kay McNulty, Betty Jennings, Betty Snyder, Marlyn Wescoff, Fran Bilas, and Ruth Lichterman—programmed one of the first electronic computers, transforming defense applications. Together, these women's innovation and leadership shaped technology in defense and continue to inspire future generations.

1 Introduction

Women have been an integral part of computer science from the very beginning starting with Augusta Ada King, Countess of Lovelace, otherwise known as Ada Lovelace. The daughter of the poet, Lord Byron, collaborated with Charles Babbage to develop the Analytical Engine and presents the case that machines can be used to perform complex mathematical functions. She is widely celebrated as the first computer programmer. Lovelace translated a paper in French written by an engineer who attended Babbage's lecture about the Analytical Engine at the University of Turin. In her translation into English, she added extensive notes and "described a sequence of steps that could be used to solve mathematical problems. It was essentially the first computer program" [1]. She later published these notes, which were three times longer than the original paper, in 1843.

Lovelace, seen in Fig. 1, has inspired many in the computer science field. In the United Kingdom, the second Tuesday of every October is designated as Ada Lovelace

T.-H. N. Schnell (✉)
Rotary Mission and Systems, Lockheed Martin, Moorestown, NJ, USA
e-mail: hanhschnell@gmail.com

A. Weissman (ed.), *Women in Advanced Computing and Software for Defense*, Women in Engineering and Science, https://doi.org/10.1007/978-3-032-17466-6_2

Day to commemorate her achievements and those of other women in science, technology, engineering, and math (STEM) [2]. Since 1994, the Grace Hopper Celebration (GHC) annually brings together thousands of women from all over the world to learn, network, and celebrate their achievements. Founders, Dr. Anita Borg and Dr. Telle Whitney, created GHC to honor the legacy of Grace Hopper and to provide women in technology a place for learning and career growth [3].

Fig. 1 Portrait of Lovelace by Margaret Sarah Carpenter, 1836 [4]

In the 1970s, the U.S. Department of Defense commissioned the creation of a new programming language focusing on reliability, safety, and security. That new language was officially named Ada to honor Ada Lovelace [5]. Coincidentally, this author used the Ada programming language to develop software for the U.S. Navy. The software that the author developed performed readiness and health-status monitoring of the hardware and software systems within the Aegis Combat System on board the Aegis cruiser and destroyer ship platforms.

In her research to write this chapter, she has learned more about the role of women in software than she did taking engineering and software programming classes in college. During her college years in the late 1980s and early 1990s, she only had one engineering professor who was a woman. This chapter will showcase several women who played pivotal roles in the advancement of computing and software in defense and show how their contributions made incredible impacts on computer science. It is the author's hope that these stories will inspire others as it has inspired her to learn more about the history of women in technology.

2 Rear Admiral Grace Murray Hopper, Ph.D.

One of the most notable figures in computer science is Rear Admiral Grace Murray Hopper, U.S. Navy (Ph.D.). While Ada Lovelace is considered the first computer programmer, Hopper is known as the creator of the first compiler that translates English terminology into machine code that computers can interpret and act on those commands (Fig. 2).

Grace Murray was born on December 9, 1906, in New York City, but attended school in New Jersey. After graduating Phi Beta Kappa from Vassar College in 1928, with a bachelor's degree in mathematics and physics, she continued her postgraduate education at Yale University where she earned a master's degree in mathematics. After graduation in 1930, she returned to Vassar College to teach mathematics while simultaneously pursuing her Ph.D. in Mathematics, which she completed in 1934. It was during this time that she met and married Vincent Hopper and took on his last name, but they divorced in 1945 [7].

In 1941, Hopper took a one-year sabbatical to study with the famous mathematician Richard Courant at New York University on a Vassar Faculty Fellow. While at Vassar, she also incorporated the updated research in physics to transform the mathematics curriculum, making it less theoretical and with more real-world application.

With the United States entering World War II, Hopper felt compelled to serve her country. She convinced Vassar to grant her permission to leave her teaching position and join the Women Accepted for Volunteer Emergency Service (WAVES), a branch of the U.S. Navy Reserve [8]. She attended training at the Naval Reserve Midshipmen's School at Smith College in Northampton, Massachusetts. The WAVES went through the same courses as the men to become naval officers, which was normally 3 months of training. However, the women learned much more quickly,

Fig. 2 Rear Admiral Grace Murray Hopper (Courtesy of U.S. Navy) [6]

and the training was reduced eventually to only 1 month. Upon their graduation from the program, they would be commissioned as either ensigns or lieutenants junior grade depending upon their prior work experience and their performance in the program [9].

Since Hopper graduated first in her class, she was commissioned as a lieutenant, junior grade. She was then assigned to the Bureau of Ships Computation Project at Harvard University. She joined the Mark I computer programming staff headed by Howard H. Aiken. It was in this environment that Hopper began to make an indelible mark on the computer science world [10].

While working on the Mark I team, Hopper and Aiken co-authored papers on the system. With information about how to use and maintain the Mark I being documented in many disparate papers; Hopper was given the task to compile these sources and consolidate the information into a single book called "A Manual of Operation Automatic Sequence Controlled Calculator." This book, which documented both the hardware and the manner and how to program the machine, became the very first textbook on computer programming [11].

At the end of World War II, Hopper continued to serve in the Navy Reserve after her request to join the regular U.S. Navy was declined due to her age. She was 38 years old. She turned down a full professorship at Vassar to work as a research fellow at Harvard under a Navy contract. During this time, the team developed the Mark II computer. In 1947, she discovered a moth stuck in one of the computer's relays causing problems with the computer's function. While she was not the first to use the term, Hopper is credited with popularizing the use of "debugging" a computer malfunction.

One of the issues that Hopper wanted to address was the tedious, repetitive, and error-prone nature of some of the work that the team had to do to program the computer. She had collected pieces of code that performed certain functions that she previously validated that they worked properly. However, whenever she wrote a new program, those pieces of code had to be copied manually to every computer address. To solve this problem, she created a library with each code set being assigned a unique name. Using this library, she was able to "tell the computer to put them together, copy them, and add them to the addresses" [12]. This code became the first compiler, and it translated mathematical notation into machine language. Hopper presented her findings in a paper. While this innovation was timesaving and produced quality code, it took 2 years before her colleagues would accept the concept of a compiler. In her words, "They had to because it worked" [12]. She stayed at the Harvard until 1949.

After leaving Harvard, Hopper joined UNIVAC where she worked until 1967. She recognized that while the compiler she created was able to translate mathematical equations into machine language, she wanted to develop a mechanism to translate human language into machine language. Hopper said, "So I decided data processors ought to be able to write their programs in English, and the computers would translate them into machine code" [12]. She created FLOW-MATIC which was the first compiler for data processing [13].

Recognizing the need to standardize how such a programming language should be developed with a set of standards and guidelines, the Pentagon created the Conference on Data Systems Languages (CODASYL) in 1959. The intent for this organization was to bring together the top technical experts and program leaders in the department of defense to develop the standards by which the coding languages need to conform and the requirements that must be met to increase portability between different computing hardware systems. Hopper was a member of CODASYL [14]. As a result of this conference, a committee was formed to create this new programming language called COBOL (*Co*mmon *B*usiness-*O*riented *L*anguage). Hopper served on this committee and directed the development of this programming language. In December 1960, the team proved the language's portability and compatibility with multiple computer systems by running the same code on both the UNIVAC and the RCA computers. Thus, she became known as the Mother of COBOL [15]. Its use proliferated throughout the various government departments. It also became heavily used in the financial and banking industries due to the data-intensive nature of their processing needs.

Hopper continued to serve on CODASYL that monitored COBOL. She retired from the U.S. Navy Reserves in 1966 with the rank of commander but was recalled to active duty in 1967, to help standardize the U.S. Navy's computer languages. In 1983, she was promoted to the rank of commodore (see Fig. 3) and ultimately became a rear admiral in 1985. She was the oldest officer in active duty with the U.S. Navy when she retired in 1986, at the age of 79 [16]. She then became a senior consultant with the Digital Equipment Company until her death in 1992.

The Society of Women Engineers (SWE) was founded in 1950 and Hopper was one of the organization's founding members. She was a role model and voice for women engineers throughout her career. SWE established a scholarship named the Admiral Grace Murray Hopper Scholarship in 1992 to honor the woman known as "Amazing Grace." Two years later, in 1994, Hopper received SWE's Achievement Award, the highest award given by the organization [18].

Hopper has been recognized by many different organizations within her lifetime. She was "elected a fellow of the Institute of Electrical and Electronic Engineers

Fig. 3 Hopper is sworn in by President Ronald Reagan as commodore in 1983 (Image courtesy of U.S. Navy) [17]

(1962) and was named the first computer science Man of the Year by the Data Processing Management Association (1969)" [18]. The Computer History Museum inducted Hopper as a Computer History Museum (CHM) Fellow in 1987 [19]. She was awarded the Defense Distinguished Service Medal, the highest non-combat award by the Department of Defense. The U.S. Navy named an Arleigh Burke-Class guided-missile destroyer, DDG-70, the USS Hopper pictured in Fig. 4, in her honor [20]. She received the National Medal of Technology in 1991, the first individual woman to receive this honor. On November 16, 2016, President Barack Obama bestowed the Presidential Medal of Freedom, the Nation's highest civilian honor, on Hopper (posthumous) for her pioneering work that enabled today's digital world [21].

Rear Admiral Grace Murray Hopper, Ph.D., was a legendary figure in the history of computing. She broke barriers and brought new ways of thinking that helped to push the boundaries of innovation. She was a pioneer, a visionary, and a teacher who ushered in the computer world that we have today. She spoke about science and technology in terms that most people can grasp and understand. She also showed us that gender and age were not limits in the pursuit of a career in STEM. Hopper died on January 1, 1992, at the age of 86 and was interred in Arlington National Cemetery.

Fig. 4 DDG-70 USS Hopper [22]

3 The ENIAC Six

While Harvard University was developing the Mark I electromechanical computer with the U.S. Navy, the Moore School of Electrical Engineering at the University of Pennsylvania was looking to advance computing with a computer of their own with the U.S. Army. They secured a contract with the U.S. Army to build a new computer using vacuum tube technology. The ENIAC (Electronic Numerical Integrator and Computer) became the first electronic computer in the world.

During World War II, there was an urgent need to update the ballistic firing tables so that soldiers in the field could accurately point artillery toward their intended targets. Many factors impact the ballistic trajectory of fired ordnance. The U.S. Army's Ballistic Research Laboratory (BRL), a research facility under the U.S. Army Ordnance Corps, studied the effects of these factors including wind and atmospheric conditions. They partnered with the Moore School to host a department of "computers" at the university. The U.S. Army sent officers from the Aberdeen Proving Grounds as liaisons at the school. During that time, mathematicians who performed complex calculations by hand were called "computers." The U.S. Army hired about 200 computers to manually calculate ballistic trajectories using differential equations to work at the Moore School. Many of them were women. Captain Herman Heine Goldstine, who had a doctorate in mathematics, was the supervisor of the "computers" team [23].

When a new project to build the first electronic computer needed people to program it, Goldstine chose his top six computers to staff it. Their names were Kathleen "Kay" McNulty (1921–2006; her married names were Mauchly and Antonelli), Frances Bilas (1922–2012; her married name was Spence), Jean Jennings (1924–2011; her married name was Bartik), Ruth Lichterman (1924–1986; her married name was Teitelbaum), Marlyn Wescoff (1922–2008; her married name was Meltzer), and Frances "Betty" Snyder (1917–2001; her married name was Holberton). Although John Mauchly and J. Presper Eckert invented and built the ENIAC, these women brought the computer to life and made it useful. These women will come to be known as the ENIAC Six [24].

Kathleen Antonelli and Frances Spence both earned a degree in mathematics from Chestnut Hill College in 1942. They were both hired as computers at the Moore School when they answered an advertisement for women math majors and knew both math and physics. Jean Bartik was the only mathematics major to graduate from Northwest Missouri State University in 1942. Betty Holberton, who was from Philadelphia, graduated from the University of Pennsylvania with a degree in journalism in 1939, and would eventually become one of ENIAC's co-lead programmers. Marlyn Meltzer was also from Philadelphia and graduated from Temple University with a degree in social studies in 1942. She initially worked on weather calculations because she knew how to operate an adding machine then performed ballistics calculations. Ruth Teitelbaum was from Far Rockaway Beach, New York, graduated from Hunter College with a degree in mathematics and joined to the Moore School computers to work on ballistics calculations [25].

What makes accomplishments of these six women truly remarkable was that they had to determine how to program the computer initially from reviewing blueprints of the computer and conversations with the engineers on the project because the U.S. Army did not initially grant them security clearance to see the physical hardware [26]. This is a daunting task as the ENIAC was comprised of "17,468 vacuum tubes, 7,200 crystal diodes, 1,500 relays, 70,000 resistors, 10,000 capacitors and around 5 million hand-soldered joints" [27]. With no instruction manuals or programming guides to follow, the ENIAC Six had to carefully study these drawings to determine what the machine was capable of and not capable of doing. This knowledge was extremely useful in determining their approach to programming this computer. It would be months later before they were able to obtain security clearances to be on the physical hardware.

The six programmers worked in pairs on different aspects of the programming content. This concept of equal partners programming together is known today as pair programming. The Agile Alliance defines pair programming as "two programmers sharing a single workstation (one screen, keyboard, and mouse among the pair). The programmer at the keyboard is usually called the "driver," the other, also actively involved in the programming task but focusing more on overall direction is the "navigator"; it is expected that the programmers swap roles every few minutes or so" [28]. The ENIAC Six were the original pair programmers. They were the screen, keyboard, and mouse all in one.

Meltzer and Teitelbaum helped prepare the ballistics program. Frances Bilas Spence and Kathleen "Kay" McNulty Mauchly Antonelli used the ENIAC to calculate the ballistics equations. Jean Jennings Bartik and Betty Holberton handled the ENIAC's Master Programmer that controlled all program sequences of the ENIAC [29]. When the ENIAC's logical circuits lacked the capacity to calculate some trajectories, Kay's innovation was to command the Master Programmer to repeat code. Thus was born the subroutine. The ENIAC Six used subroutines and nested subroutines to modularize their code and maximize the capability of the ENIAC [30]. They led the programming of the ballistics trajectory for the successful February 15, 1946, demonstration. Sadly, none of these women were invited to the demonstration nor the celebration that followed it. Their contributions were not acknowledged to the public [24].

Fortunately, Kathy Kleiman saw the photograph of the ENIAC with some women in the photo shown in Fig. 5 while she was an undergraduate computer science student at Harvard University. When she inquired about the women, she was told that they were models intended to make the machine look better like the "refrigerator ladies" in printed advertisements. Undeterred, Kathy did further research and found one reference that named all six women. She then initiated the decades-long ENIAC Programmer Project to locate and interview all six women to tell the world their story. She first created a documentary film called *The Computers: The Remarkable Story of the ENIAC Programmers* [31] which won the Best Documentary Short at the 2016 United Nations Association Film Festival. She later wrote a book titled *Proving Ground: The Untold Story of the Six Women Who Programmed the World's First Modern Computer* [32] that shared more in-depth details of her research [33].

Fig. 5 Betty Jennings (left) and Frances Bilas (right) operating the ENIAC's main control panel while it is located at the Moore School. (Image courtesy of U.S. Army)

Kleiman also lobbied for these women to be recognized at the 50th Anniversary of ENIAC in 1996. All six ENIAC programmers were inducted into the Women in Technology International Hall of Fame in 1997 [29].

The ENIAC story is not complete without mention of another woman, Adele Katz Goldstine (1920–1964). She was a mathematician who graduated with a master's degree in mathematics from the University of Michigan. It was there that she met and married Captain Herman Heine Goldstine in 1941. She came to Philadelphia with her husband when he was assigned to the Moore School at the University of Pennsylvania. Adele served as a mathematics instructor for the women "computers" at the Moore School. She trained some of the original programmers of ENIAC on how to manually calculate ballistic trajectories using complex differential calculations. For the ENIAC project, her task was to document all the work that the ENIAC Six innovated and create the technical description of the ENIAC. In other words, she wrote the computer's first operator's manual [34].

When the ENIAC moved to Aberdeen Proving Grounds in Maryland, some continued to work on the project and moved to Maryland. Three of the programmers moved to Maryland. The remaining three chose to remain in Philadelphia.

While working on the ENIAC, Frances Bilas met Private First-Class Homer Spence, an electrical engineer who was also assigned to the project. They married in 1947 and moved to Maryland when the ENIAC moved. Homer became the Chief of the Computer Research Branch. Frances continued to work on the ENIAC until the birth of their first child. Subsequently, she chose to stay home and focus on being a mother. She raised their three sons and never returned to the workforce [35].

Kathleen "Kay" McNulty moved to Maryland to continue her work with ENIAC. John Mauchly and Presper Eckert started their own company, Eckert-Mauchly Computer Corporation in 1947. The following year, she married John Mauchly and moved back to the Philadelphia area to work at this new company. John was a widower with two children. Eventually, they had five more children.

Kay developed the software design that became the foundation for the Binary Automatic Computer (BINAC) and UNIVersal Automatic Computer I (UNIVAC I) computers while John designed the hardware. After the ENIAC Six were recognized, she spoke at various computer-related events, even being a keynote speaker at the 40th anniversary event. At first, she spoke of her husband's work and her own only in passing. However, after her own efforts were made publicly known, she spoke about her own incredible contributions [36].

Ruth Lichterman moved with the ENIAC and continued her work at the Ballistics Research Laboratory at Aberdeen, Maryland. She worked there for two more years teaching others how to program and use the tools she had developed for the ENIAC [37]. Ruth was the last of the original six to remain working at the Ballistics Research Laboratory. She retired in 1948 after marrying Adolph Teitelbaum.

The other three women, not wanting to move to rural Maryland, pursued other paths. Frances Elizabeth "Betty" Snyder Holberton, wanting to stay in Philadelphia, joined Eckert-Mauchly Computer Corporation. She continued to innovate. For the UNIVAC I, she created a programming language that used mnemonic characters to signify logic. This language became known as C-10, and she published several papers to describe its use [38].

Betty also designed the control panel to have the numeric keypad next to the alphanumeric keyboard. It was her suggestion to change the UNIVAC's paint color from black to the gray color that many are familiar with seeing today. She then worked at David Taylor and the National Bureau of Standards (now known as the National Institute of Standards and Technology). She served on committees that created COBOL (Common Business-Oriented Language) and wrote standards for FORTRAN (Formula Translation) computer languages. These committees also set other national and international computer standards. She retired in 1983 [39].

Jean Jennings met William John Bartik, an engineer, while she was working at the Moore School. They were married in 1946 with John Mauchly walking her down the aisle. She led a team to develop the capability for ENIAC to store memory on punch cards, making programming easier.

Jean joined Eckert and Mauchly's new company to stay in the Philadelphia area. She worked with them to design the BINAC. It introduced the ability for a computer to store data on magnetic tape instead of punch cards. Next, she developed the logical design for UNIVAC I working with Frances Holberton. It became the first non-military computer when the Census Bureau purchased it in 1951. Remington Rand Corporation purchased Eckert-Mauchly Computer Corporation and made it the UNIVAC division of the company.

When her husband got a job in Washington, D.C., Jean transferred to that location. She trained "Census Bureau programmers, worked with the salesmen selling Remington Rand equipment, and worked on programs for the Aviation Supply Office for the Navy on the UNIVAC" [39]. But when William Bartik joined Remington Rand at the Philadelphia location in 1951, she resigned because the company did not allow for both husbands and wives to work in the location. They started a family, and she spent the next 16 years raising their three children. In 1967, Jean earned her master's degree from the University of Pennsylvania and returned to work in computer science. She worked at a variety of companies such as Honeywell and Data Decisions until she retired in 1986. In addition to being inducted into the Women in Technology International Hall of Fame with her fellow ENIAC programmers, Jean also received numerous other technology awards including the prestigious Computer Pioneer Award in 2008 from the Institute of Electrical and Electronics Engineers' Computer Society Board of Governors for her pioneering work with ENIAC, BINAC, and UNIVAC I [40].

Marlyn Wescoff met Phillip Meltzer and were married in 1947. She resigned from the ENIAC project soon afterwards. Marlyn spent the remainder of her life as a volunteer reading to children at the Shir Ami Library, delivering food for "Meals on Wheels," and knitting hats for chemotherapy patients [41].

In 1981, the Institute of Electrical and Electronics Engineers (IEEE) Computer Society Board of Governors established the Computer Pioneer Award in Honor of the Women of ENIAC Award. This award recognizes and honors "the vision of those people whose efforts resulted in the creation or expansion and continued vitality of the computer industry. The award is presented to outstanding individuals whose main contribution to the concepts and development of the computer field was made at least 15 years earlier" [42]. Among the charter recipients of this IEEE award is Rear Admiral Grace Murray Hopper.

The ENIAC Six taught us that not only did women belong in programming and computer science, but women were also the front runners in the shaping of the technology. They were the leaders who innovated and taught others to follow in their path. Their efforts were obscured for decades before another woman in computer science who was in search of "people who looked like her" discovered their incredible contributions and made it known to all. This lesson further emphasizes the importance of documenting the accomplishments of these unsung heroines who were so vital to the advancement of computers and technology. It is through these stories that we promote STEM careers and encourage the next generation of innovators and pioneers.

4 Margaret H. Hamilton

The ENIAC Six showed us that women have always been in the forefront of programming. Rear Admiral Grace Murray Hopper's compiler made programming accessible to many through its translation of everyday language to code that the computers can execute. Margaret H. Hamilton showed the world that software development required engineering to achieve expected results and coined the term "software engineering." She led the development of the guidance and control systems for both the inflight command and lunar modules. Her work enabled the successful Apollo missions to the moon (Fig. 6).

Margaret Heafield was born in Paoli, Indiana, on August 17, 1936 [44]. Her mother was a high school teacher, and her father was a college professor. They moved the family to different parts of Indiana, Ohio, and Michigan following the work. She ultimately lived in the Upper Peninsula of Michigan where she attended high school. When she was 16 years old, she started to work at the Arcadian Copper Mine running tours and hiring staff, eventually also handling the finances for the company. This experience had a great influence in her life. She learned a strong work ethic and working hard to earn everything that she got [45].

Fig. 6 Hamilton pictured in 1989. (Image courtesy of NASA) [43]

Hamilton graduated from high school in 1954 [46]. She was accepted in several colleges but chose to attend the University of Michigan to study mathematics. She felt that the college provided the most real-world environment and offered her a good scholarship. However, she transferred to Earlham College after one semester because it was the school that her mother and many of her family members had attended. Because she had moved around so often as a child, this quick change did not faze her. At Earlham, she studied mathematics and philosophy, graduating in 1958 with a Bachelor of Arts in mathematics. It was also at Earlham that she met and married James Hamilton [44]. Since James still had one more year of college to complete, Hamilton took a position at a high school and taught mathematics and French until he finished his studies [45].

In 1959, after James graduated from Earlham, he was accepted to Harvard Law School, and the couple moved to Boston, Massachusetts. Although Hamilton had been accepted to Brandeis University to study abstract mathematics, she ultimately chose to accept an offer to work at the Massachusetts Institute of Technology (MIT) [44]. At MIT, Hamilton was introduced to software programming for the first time. She developed weather prediction software on the Librascope LGP-30 and the Digital Equipment Company (DEC) PDP-1 computers, to model the meteorological effects of global climate changes for Professor Edward N. Lorenz [46]. After giving birth to her daughter, Lauren, that November, she returned to MIT to continue her work [47].

Although she enjoyed the work at MIT, Hamilton needed a job that had better pay so she began to look for other opportunities. When Lincoln Laboratory advertised that they were hiring programmers, she immediately contacted them for an interview. Interestingly, as the recruiter usually interviewed male candidates in his hotel room, he had to pivot and interviewed Hamilton at the hotel bar. Naturally, she was hired and began to work on the AN/FSQ-7 XD-1 computer [45].

Hamilton worked with the Semi-Automatic Ground Environment (SAGE) system at MIT Lincoln Laboratory, where she wrote software to differentiate aircraft radar signatures from electronic noise. It was the United States' first air defense system developed during the Cold War [48].

While working on SAGE, she discovered that many programmers often tried to write "tricky" code as she puts it, with no comments so that others would not know how it worked. One of the other programmers had employed this "tricky programming" and then left Lincoln Laboratory. One of Hamilton's first assignments was to figure out his programming so that she could make changes to it. She was able to get the program to print out onto the printer that no one else had been able to do before. She learned the importance of comments in the software and added comments to that code. Hamilton said that "I could forget my own code, because I've been tricky. Now I'm going to trick myself" [45]. She made sure that any code that she wrote or worked with had proper comments. Her example influenced others to also write more documentation in their code.

Hamilton also learned about software reliability because an unsuccessful program on the SAGE computer would result in bells and clangs to alert you to a problem, and everyone would know whose code caused it. The operators would also run the programs 24 h a day. The art of software programming was being invented for SAGE to make it more efficient and reliable. New concepts like subroutines and central service subprograms were also introduced [49]. She recalled a time when she received a phone call at 4 a.m. because there was something wrong with her program and had to come in to fix it for them [45].

When Hamilton learned that MIT Instrumentation Lab, which later became Draper Laboratory, was hiring programmers to work on the Apollo Missions, she eagerly applied. She was offered two positions, one in software support systems and one in flight software for the onboard computers. Although she really wanted to work on the flight software, she did not want to offend anyone and asked them to flip a coin. Fortunately, onboard flight software won the coin flip [45]. She took all the skills and innovation that she learned working on the SAGE project and applied them to this new assignment.

Since software development was so very new, there was a great deal of creation of process and product. Hamilton led the Instrumentation Lab's Software Engineering Division. There was a lot of engineering in the process, so she used the term software engineering to describe the work that she and her team were doing. Her team developed the software that ran on the Apollo Guidance Computer (AGC), which was a 1-cubic-foot example of innovative engineering. The AGC was one of the first computers to use silicon integrated circuit chips in a significant manner and catapulted the microchip technology. During a time when computers were housed in large rooms, its compact size was quite a feat of innovation. The AGC's 36 kilobytes of read-only memory and 2 kilobytes of random-access memory were used in a very sophisticated manner to achieve the expected outcome [50] (Fig. 7).

Hamilton knew that rigor and intentionality were required to develop reliable software especially when it was mission and safety critical. "One of the most complex tasks required to make the lunar landing possible was the unprecedented feat of navigation required to guide a craft through a round-trip of more than a half-million miles, and to achieve a controlled, precise landing on the lunar surface on the very first try—there was no fuel available for a wave-off and another attempt" [52]. With such high stakes, the code had to perform flawlessly.

The code that Hamilton wrote for the AGC was so extensive that when printed out and stacked in a pile, the stack of code was taller than her height which can be seen in Fig. 8. This image was so iconic that Lego made a replica of this photo as part of their Women of NASA Lego Set [53]. She is also a featured Non-Player Character in the CHM Minecraft World where she is shown standing beneath the Lunar Module [54].

To validate the code, the Software Engineering Division used simulators to test the AGC code in a representative environment. Hamilton worked with these simulators. From when her daugher, Lauren, was a baby, Hamilton would take Lauren to work when she worked at night or on weekends at MIT under Professor Lorenz since her daughter would be mostly sleeping. So, bringing her daughter to work was a very natural action for her on nights and weekends when she worked on Apollo 8. Hamilton recalled a story of when Lauren played with the hardware simulator, pushing buttons like she saw her mother doing and caused the whole system to

Fig. 7 Apollo Guidance Computer (AGC) and Navigation System [51]

crash. Her daughter had selected the prelaunch program (P01) when the spacecraft was already in flight. Hamilton called it the "Lauren Bug." Although she urged the mission leaders to allow her to add protective code that would prevent execution of prelaunch functions during flight and give feedback to the astronauts, she was told that astronauts would not make that mistake due to their training. Still concerned, she successfully pushed for adding a program note so that it would indicate the source of the issue if the situation did occur [46].

On Day 4 of the Apollo 8 Mission, during the 8th orbit around the moon, P01 was accidently initiated, and the crew had to re-align the guidance platform [56]. Hamilton was at MIT Lincoln Lab in the Switching, Conference, and Monitoring Arrangement (SCAMA) facility used during the Apollo Missions when it happened. Poring through the code listings to determine what happened, Hamilton "said, 'It's the Lauren bug. I know it's the Lauren bug. This is exactly what happened.'" After that incident, Hamilton was allowed to add the preventive code [46].

Thinking through "what if" scenarios was a way for Hamilton to determine different situations that can arise during the mission and how to address them. One of those conditions that she wanted to address was the ability of the software to alert the astronauts of an emergency. She needed a way to asynchronously communicate but they could not interrupt their display where they were performing work. She worked with the hardware engineers to make more hardware available for the duration of

Fig. 8 Hamilton next to the printed code for the AGC (Image courtesy of NASA) [55]

the mission to handle emergencies. She then had to convince the system engineers that parallel processing was possible by developing an initial solution that added a priority display to alert the astronauts of an emergency and give them the option to address the issue or ignore it. This change alerted the crew of Apollo 11 that they were overloading the computer which helped them recognize that they switched on unnecessary processing for the computer that could be halted to allow for the system to perform nominally [46]. In her own words, "It quickly became clear that the software was not only informing everyone that there was a hardware-related problem, but that the software was compensating for it. With only minutes to spare, the decision was made to go for the landing. The rest is history" [57].

Hamilton continued to work on the Apollo Missions as well as the first U.S. space station, Skylab, until 1972 [58]. She left MIT Lincoln Laboratory, co-founding Higher Order Software (HOS) with MIT colleague, Saydean Zeldin. Their company worked with the National Aeronautics and Space Administration (NASA) and the U.S. Department of Defense customers [59]. In 1986, she left HOS and founded

Hamilton Technologies, Inc., where she created the Universal Systems Language (USL) and its automation tool suite. Her company provides "products and services to modernize the planning, system engineering and software development process in order to maximize reliability, lower cost and accelerate time to market" [60].

Hamilton received many honors, awards, and recognition throughout her career. In 1986, the Association of Women in Computing presented her with the Augusta Ada Lovelace Award. In 2003, NASA granted the NASA Exceptional Space Act Award to Hamilton, recognizing her contributions to software development by granting her $37,200 award—the highest prize ever given up until that time. Dr. Paul Curto, who nominated her for the award said, "I was surprised to discover she was never formally recognized for her groundbreaking work. Her concepts of asynchronous software, priority scheduling, end-to-end testing, and man-in-the-loop decision capability, such as priority displays, became the foundation for ultra-reliable software design" [61].

In 2016, President Barack Obama presented her with the Medal of Freedom for her contributions to software engineering and space exploration as shown in Fig. 9. Coincidently, Hopper was also honored with a posthumous Medal of Freedom on the same day. CHM awarded Hamilton the Fellow Award in 2017 and recorded an extensive interview with her [48]. In 2019, the Intrepid Museum honored her with the Intrepid Lifetime Achievement Award for her work to lay the foundation for software engineering [62].

Margaret H. Hamilton is a pioneer in software engineering, demonstrating that the software field was engineering and required design, rigor, and discipline. She

Fig. 9 Hamilton receives the Presidential Medal of Freedom from President Obama. (Official White House Photo by Lawrence Jackson) [63]

asked questions to help others think outside of traditional mindsets and boundaries. She introduced safety and reliability into software by exploring "what if" scenarios and including preventive code. She incorporated lessons learned into software development. She strove to create a culture where errors, once discovered, would not be repeated because all documented anomalies were sent to the development team for awareness and resolution. Her focus on software development as an engineering effort resulted in the discipline needed to ensure that safety–critical software required to support crewed space flight always supported their missions. Her methods and concepts laid the foundation for modern software engineering.

5 Elizabeth Jocelyn "Jake" Feinler

Margaret Hamilton helped to get humankind to the moon. But it was Elizabeth "Jake" Feinler that brought the people of the world closer together by laying the foundation for the internet. She is a computer information scientist whose pioneering work helped to define the internet architecture and its organization. She worked at the Stanford Research Institute (now called SRI International) for decades. She became the Director of the Network Information Systems Center, where she managed the Network Information Center (NIC) for ARPANET, Defense Data Network (DDN), and drove standardization that allowed the Internet to easily grow and adapt and known as the Mother of WHOIS (Fig. 10).

Feinler was born in Wheeling, West Virginia, on March 2, 1931. Her parents intended to call her "Betty Jo" but her 2-year-old older sister said what sounded like "Baby Jake." Her family thought it was funny, and the name "Jake" stuck. She grew up in West Virginia and attended West Liberty State College, where she majored in chemistry. She was the first person from her family to go to college. She later pursued graduate work in biochemistry at Purdue University but took a "short" break in 1958 to earn some money. She worked at Chemical Abstracts Service as an assistant editor on a project to index the world's chemical compounds in the *Chemical Abstracts 5th Decennial Index*. That break became a permanent one, and she never went back [65].

In 1970, Feinler then joined SRI, originally to work in the Literature Research Division to help develop the *Handbook of Psychopharmacology* and the *Chemical Process Economics Handbook*. According to a 2002 interview with Janet Abbate of the Institute of Electrical and Electronics Engineers (IEEE) History Center, when she was bored, Feinler would visit Doug Engelbart's team because she was interested in their work. She told him that she was interested in working on the project. In 1972, Doug asked her to join the project. Feinler joined the Augmentation Research Center Group and worked on the *Resource Handbook for the Internet*. In creating this handbook, she asked many insightful questions to learn everything about ARPANET (Advanced Research Projects Agency Network). Because she learned its many details and various aspects, she became a subject matter expert and a technical resource for everyone. A year later, she became the principal investigator for the ARPANET Network Information Center (NIC) [66].

Fig. 10 Elizabeth "Jake" Feinler [64]

The first connection on this network occurred in fall 1969 with a user on a computer at the University of California in Los Angeles connecting with a computer at SRI using the Interface Message Processor (IMP) program. Two weeks later, computers at the University of California in Santa Barbara and the University of Utah joined the network. The network quickly expanded with additional nodes the following year to include U.S. military bases, government laboratories, and more universities [67].

Feinler's team developed the first white pages and yellow pages of the internet. Feinler recognized that everyone on using the system had different and sometimes opposing points of view. Therefore, when she developed the directories, she ensured that it was generic by removing their titles so that people can collaborate and communicate without bias.

Feinler played a critical role in the transition of the internet with a domain naming system. Her vision of separate top-level domain spaces for different types of users by assigning designations such as.com,'.gov,.mil,.edu,.net, and.org was forward thinking and allowed for it to grow while still providing the spaces to be separate and unique. These domain naming systems are still used today. She also managed the Host Naming Registry. She created the WHOIS system that allowed users to search for domain names and internet protocol (IP) addresses across the system. Therefore, Feinler became known as the Mother of WHOIS [68].

While at SRI, she embraced her role as a trailblazer for other women in technology and used her position to invite other women to support her projects. She intentionally made opportunities available for other women and became their mentor to ensure their success. Her philosophy of bringing others with her as she rose was present throughout her career [66].

In 1989, Feinler left SRI to work at NASA's Ames Research Center and retired in 1996. She then focused her energy on capturing the history of the formation of the internet and published a history of the NIC in 2010. She volunteered at the Computer History Museum and donated her own notes and materials into the museum's archives. She also co-authored a timeline on electronic mail. For her crucial role in the formation of the internet, Feinler was inducted into the SRI Alumni Hall of Fame in 2000 and the Internet Society Internet Hall of Fame in 2012. In 2013, she received the Internet Engineering Task Force (IETF) Jonathan B. Postel Service Award. She was inducted into the Women in Technology International Hall of Fame in 2018 [69]. The next year, she was inducted into the Consumer Technology Association Hall of Fame [70].

Elizabeth Jocelyn "Jake" Feinler was a trailblazer and role model for women in technology. She mentored others and gave them opportunities to practice engineering and technology skills. She ensured history was preserved by volunteering her time and energy to archiving and documenting how the internet was created and how electronic mail came to be. Her contributions have served to connect the world.

6 Conclusion

The pioneering women presented in this chapter were leaders in their field and trailblazers for other women in science, technology, engineering, and math. They had the vision and the drive to not only achieve their goals but also make an indelible mark on the world. When invention and outside-the-box thinking were needed, they provided their insights. Their contributions moved technology forward and provided solutions that made a difference in how we interact with the world and with each other.

These amazing women answered their nations' call to serve when it was uncommon for women to do so. Their fortitude and ingenuity drove revolutionary technology changes that advanced the U.S. defense capability. They helped the warfighter in the field with situational awareness, put men on the moon, connected researchers in cyberspace, and advanced computing, making science fiction a reality. But these innovations ultimately extended to civilian use and made impacts across the globe. In a normal course of a day, millions of people benefit from the efforts of these incredible women without the knowledge or understanding of their origin.

Many people assume that technology was invented by men, implying that STEM careers were not meant for women. The women highlighted in this chapter did not allow external forces to silence their voices because of their gender. Hopefully, through their example, others will find their own passions and use these role models as guidance for their own journey to make a positive impact on the world. In addition to the trailblazing women featured in this chapter, there are numerous others who have made and continue to make significant contributions to the technological advancements that shape our world. It is essential to uncover and share the stories of these contemporary innovators, highlighting their profound impact and inspiring future generations to recognize and celebrate their achievements.

References

1. Blakemore, E. (2016). 'Five things to know about Ada Lovelace'. *Smithsonian Magazine*. Available at: https://www.smithsonianmag.com/smart-news/five-things-know-about-ada-lovelace-180960745/ [Accessed 30 Sep. 2025].
2. Bodleian Libraries. (n.d.). *About Ada Lovelace*. Available at: https://blogs.bodleian.ox.ac.uk/adalovelace/about-ada-lovelace/ (Accessed 30 September 2025).
3. AnitaB.org. (n.d.). *About Grace Hopper Celebration*. Available at: https://ghc.anitab.org/about (Accessed 30 September 2025).
4. Robins, R. (1999). Ada and the First Computer. Available at: https://www.cs.virginia.edu/~robins/Ada_and_the_First_Computer.pdf (Accessed 30 September 2025).
5. AdaCore. (n.d.). *Timeline of Ada*. Available at: https://www.adacore.com/about-ada/timeline-of-ada (Accessed 30 September 2025).
6. U.S. Navy. *Rear Admiral Grace Hopper*. Official U.S. Navy photo. Accessed October 19, 2025. https://defenseimagery.mil/imagery.html#guid=f1a07fc90bc516cbf2426f45f804117be1c69d73
7. Yale University. (n.d.). *Biography of Grace Murray Hopper*. Available at: https://president.yale.edu/biography-grace-murray-hopper (Accessed 30 September 2025).
8. New-York Historical Society. (n.d.). *Grace Hopper*. Available at: https://wams.nyhistory.org/confidence-and-crises/world-war-ii/grace-hopper/ (Accessed 30 September 2025).
9. Smith College. (n.d.). *Learning to Be Navy: Grace Hopper and the Origins of Computer Programming*. Available at: https://www.smith.edu/news-events/news/learning-be-navy (Accessed 30 September 2025).
10. Grace Hopper Program. (n.d.). *About Grace Hopper*. Available at: https://gracehopperinstitute.org/about-grace-hopper/ (Accessed 30 September 2025).
11. Harvard University. (n.d.). *Harvard-IBM Mark I Manual*. Available at: https://chsi.harvard.edu/harvard-ibm-mark-1-manual (Accessed 30 September 2025).
12. Vassar College. (2017). *Legacy of Grace Hopper*. Available at: https://www.vassar.edu/stories/2017/assets/images/170706-legacy-of-grace-hopper-hopperpdf.pdf (Accessed 30 September 2025).

13. Remington Rand Univac, a Division of Sperry Rand Corporation. FLOW-MATIC Programming System. 1957. PDF file. https://www.bitsavers.org/pdf/univac/flow-matic/FLOW-MATIC_Programming_System_1958.pdf (Accessed 30 September 2025).
14. History-Computer. (n.d.). *COBOL Guide*. Available at: https://history-computer.com/technology/cobol-guide/ (Accessed 30 September 2025).
15. FedTech Magazine. (2017). *How COBOL Became the Early Backbone of Federal Computing*. Available at: https://fedtechmagazine.com/article/2017/09/how-cobol-became-early-backbone-federal-computing (Accessed 30 September 2025).
16. Encyclopaedia Britannica. (n.d.). *Grace Hopper*. Available at: https://www.britannica.com/biography/Grace-Hopper (Accessed 30 September 2025).
17. U.S. Navy, "Reagan Honors Pioneering Computer Scientist Hopper with Promotion," photo ID 9161680, Defense Visual Information Distribution Service (DVIDS). Accessed October 19, 2025. https://www.dvidshub.net/image/9161680/reagan-honors-pioneering-computer-scientist-hopper-with-promotion
18. Society of Women Engineers. (n.d.). *75th Anniversary: Building a Legacy*. Available at: https://swe.org/75th/key-themes/75th-building-a-legacy/ (Accessed 30 September 2025).
19. Computer History Museum. (n.d.). Grace Murray Hopper Profile. Available at: https://computerhistory.org/profile/grace-murray-hopper/ (Accessed 30 September 2025).
20. U.S. Navy. (n.d.). Grace Hopper Biography. Available at: https://www.history.navy.mil/content/history/nhhc/research/histories/biographies-list/bios-h/hopper-grace.html (Accessed 30 September 2025).
21. White House Archives. (2016). Celebrating Presidential Medal of Freedom Winners in Science and Technology: Garwin, Hopper, and Others. Available at: https://obamawhitehouse.archives.gov/blog/2016/11/22/celebrating-presidential-medal-freedom-winners-science-and-tech-garwin-hopper-and (Accessed 30 September 2025).
22. U.S. Navy. "The destroyer USS Hopper (DDG 70) launches a Standard Missile-3." July 31, 2009. Photo ID 9218012. *Defense Visual Information Distribution Service (DVIDS)*. Publicly released. Accessed October 19, 2025. https://www.dvidshub.net/image/9218012/destroyer-uss-hopper-ddg-70-launches-standard-missile-3.
23. Wittenberg University. (2013). *Herman Goldstine Funding Story*. Available at: https://www.wittenberg.edu/news/2013/03-14-pi#:~:text=Herman%20Goldstine%20who%20had%20doctorate,to%20the%20army%20for%20funding (Accessed 30 September 2025).
24. TilScience. (n.d.). *The Incredible True Story of How…*. Available at: https://tilscience.substack.com/p/the-incredible-true-story-of-how (Accessed 30 September 2025).
25. University of Pennsylvania Almanac. (n.d.). *75th Anniversary of the ENIAC*. Available at: https://almanac.upenn.edu/articles/75th-anniversary-of-the-electronic-numerical-integrator-and-computer-eniac (Accessed 30 September 2025).
26. ENIAC VC. (n.d.). *Remembering the ENIAC Six*. Available at: https://www.eniac.vc/writings/remembering-the-eniac-six (Accessed 30 September 2025).
27. University of Michigan, CSE. (n.d.). *ENIAC Display History*. Available at: https://cse.engin.umich.edu/about/history/eniac-dis-play/#:~:text=The%20ENIAC%20added%20and%20subtracted,on%20loan%20to%20other%20institutions (Accessed 30 September 2025).
28. Agile Alliance. (n.d.). *Pair Programming*. Available at: https://agilealliance.org/glossary/pair-programming/ (Accessed 30 September 2025).
29. WITI Hall of Fame. (n.d.). *ENIAC Programmers: Kathleen*. Available at: https://www.witi.com/halloffame/298369/ENIAC-Programmers-Kathleen-/ (Accessed 30 September 2025).
30. Fortune. (2014). *Walter Isaacson: The Women of ENIAC*. Available at: https://fortune.com/2014/09/18/walter-isaacson-the-women-of-eniac/ (Accessed 30 September 2025).
31. Ceruzzi, P. E. (n.d.). *The Computers: The Remarkable Story of the ENIAC Programmers*. [Print].
32. McCartney, P. (n.d.). *Proving Ground: The Untold Story of the Six Women Who Programmed the World's First Modern Computer*. [Print].
33. IEEE Spectrum. (n.d.). *The Women Behind ENIAC*. Available at: https://spectrum.ieee.org/the-women-behind-eniac (Accessed 30 September 2025).

34. American Philosophical Society. (n.d.). *Goldstines and the ENIAC Six*. Available at: https://www.amphilsoc.org/blog/goldstines-and-eniac-six (Accessed 30 September 2025).
35. Spence, Frances. "Frances Spence." Engineering and Technology History Wiki. Accessed October 20, 2025. https://ethw.org/Frances_Spence.
36. St Andrews Mathematics History. (n.d.). *Biography: Antonelli*. Available at: https://mathshistory.st-andrews.ac.uk/Biographies/Antonelli/ (Accessed 30 September 2025).
37. Spence, Ruth. "Ruth Teitelbaum." *Engineering and Technology History Wiki*. Accessed October 20, 2025. https://ethw.org/Ruth_Teitelbaum
38. Hopl.info. (n.d.). *ShowLanguage: EXP 5442*. Available at: https://web.archive.org/web/20200724215517/, https://hopl.info/showlanguage.prx?exp=5442 (Accessed 30 September 2025).
39. Los Angeles Times. (2001). *Story on ENIAC Women*. Available at: https://www.latimes.com/archives/la-xpm-2001-dec-15-me-15190-story.html (Accessed 30 September 2025).
40. IEEE Computer Society. (n.d.). *Volunteering Awards: Pioneer*. Available at: https://www.computer.org/volunteering/awards/pioneer (Accessed 30 September 2025).
41. Kids Kiddle. (n.d.). *Marlyn Meltzer*. Available at: https://kids.kiddle.co/Marlyn_Meltzer#:~:text=It%20weighed%20more%20than%2030,improve%20and%20evolve%20our%20society (Accessed 30 September 2025).
42. IEEE Computer Society. (n.d.). *About Women of ENIAC*. Available at: https://www.computer.org/volunteering/awards/pioneer/about-women-of-eniac (Accessed 30 September 2025).
43. NASA. "Margaret Hamilton." Apollo 11 Lunar Surface Journal. Accessed October 19, 2025. https://history.nasa.gov/alsj/a11/a11Hamilton.html.
44. Encyclopaedia Britannica. (n.d.). *Margaret Hamilton: American Computer Scientist*. Available at: https://www.britannica.com/biography/Margaret-Hamilton-American-computer-scientist (Accessed 30 September 2025).
45. Computer History Archive. (2022). *Margaret Hamilton Arcadian Copper Mine Notes*. Available at: https://archive.computerhistory.org/resources/access/text/2022/03/102738243-05-01-acc.pdf (Accessed 30 September 2025).
46. Computer History Museum. (2019). *2017 CHM Fellow: Margaret Hamilton*. Available at: https://web.archive.org/web/20190212190351/, https://www.computerhistory.org/atchm/2017-chm-fellow-margaret-hamilton/ (Accessed 30 September 2025).
47. Times News. (n.d.). *Ned Jilton II: Margaret Hamilton Helped Make the Space Program Work*. Available at: https://www.timesnews.net/living/features/ned-jilton-ii-margaret-hamilton-helped-make-the-space-program-work/article_58fdacd8-f920-11ef-8684-6f6a61d4519e.html (Accessed 30 September 2025).
48. MIT Lincoln Labs. (n.d.). *Margaret Hamilton Reflects on Life, Career, and SAGE*. Available at: https://www.ll.mit.edu/news/margaret-hamilton-reflects-life-career-and-sage (Accessed 30 September 2025).
49. MIT Lincoln Labs. (n.d.). *History of SAGE: Semi-Automatic Ground Environment Air Defense System*. Available at: https://www.ll.mit.edu/about/history/sage-semi-automatic-ground-environment-air-defense-system (Accessed 30 September 2025).
50. MIT News. (2019). *Behind the Scenes of the Apollo Mission*. Available at: https://news.mit.edu/2019/behind-scenes-apollo-mission-0718 (Accessed 30 September 2025).
51. Jurvetson, Steve. "Apollo Guidance Computer (AGC) and Navigation System from the second Lunar Module, LM-2." Photograph, July 18 2019 (uploaded October 10 2019). Flickr. License: Some Rights Reserved (Creative Commons). https://www.flickr.com/photos/jurvetson/48877874981/ (accessed October 19 2025).
52. MIT News. (2009). *Apollo TT0603*. Available at: https://news.mit.edu/2009/apollo-tt0603 (Accessed 30 September 2025).
53. LEGO. (n.d.). *Women of NASA Set*. Available at: https://www.lego.com/en-us/product/women-of-nasa-21312 (Accessed 30 September 2025).
54. Computer History Museum. (n.d.). *Making the Great Tech Story*. Available at: https://computerhistory.org/blog/making-the-great-tech-story/ (Accessed 30 September 2025).
55. MIT Instrumentation Laboratory. *Margaret Hamilton standing next to a stack of Apollo Guidance Computer source code listings*. Photograph, 1969. Accessed October 19, 2025. https://science.nasa.gov/people/margaret-hamilton/

56. NASA. (n.d.). *Apollo 8: Day 4, Orbit 7*. Available at: https://www.nasa.gov/history/afj/ap08fj/19day4_orbit7.html (Accessed 30 September 2025).
57. MIT News. (2009). *Apollo Vignette 0717*. Available at: https://news.mit.edu/2009/apollo-vign-0717 (Accessed 30 September 2025).
58. Smithsonian Magazine. (n.d.). *Margaret Hamilton Led NASA Software Team*. Available at: https://www.smithsonianmag.com/smithsonian-institution/margaret-hamilton-led-nasa-software-team-landed-astronauts-moon-180971575/ (Accessed 30 September 2025).
59. Astronomy.com. (n.d.). *Saydean Zeldin: Guiding Apollo's Engines*. Available at: https://www.astronomy.com/space-exploration/saydean-zeldin-guiding-apollos-engines/ (Accessed 30 September 2025).
60. Hamilton Technologies, Inc. (n.d.). *HTI USL Suite*. Available at: http://www.htius.com/ (Accessed 30 September 2025).
61. NASA. (n.d.). *Apollo 11: Margaret Hamilton*. Available at: https://www.nasa.gov/history/alsj/a11/a11Hamilton.html (Accessed 27 October 2025).
62. Intrepid Museum. (2019). *Salute to Freedom Gala Archive*. Available at: https://web.archive.org/web/20190924162857/, https://support.intrepidmuseum.org/campaign/salute-to-freedom-gala/c216914 (Accessed 30 September 2025)
63. Jackson, Lawrence. *President Obama presenting the Presidential Medal of Freedom to Margaret Hamilton*. Official White House Photo. November 22, 2016. Accessed October 19, 2025. https://obamawhitehouse.archives.gov/blog/2016/11/22/celebrating-presidential-medal-freedom-winners-science-and-tech-garwin-hopper-and.
64. *Elizabeth "Jake" Feinler. Photo (2011) by Elizabeth Feinler — used under CC BY-SA 3.0 (*https://creativecommons.org/licenses/by-sa/3.0/*). Source: Wikimedia Commons, File:ElizabethFeinler-2011.jpg.*
65. Internet Hall of Fame. (n.d.). *Official Biography: Elizabeth Feinler*. Available at: https://www.internethalloffame.org/official-biography-elizabeth-feinler/ (Accessed 30 September 2025).
66. Engineering and Technology History Wiki. (n.d.). *Oral History: Elizabeth "Jake" Feinler*. Available at: https://ethw.org/Oral-History:Elizabeth_%22Jake%22_Feinler (Accessed 30 September 2025).
67. DARPA. (n.d.). *ARPANET Features*. Available at: https://www.darpa.mil/news/features/arpanet (Accessed 30 September 2025).
68. SRI International. (n.d.). *75 Years of Innovation: ARPANET*. Available at: https://www.sri.com/75-years-of-innovation/75-years-of-innovation-arpanet/ (Accessed 30 September 2025).
69. WITI Hall of Fame. (n.d.). *Elizabeth Feinler: Internet Pioneer, Active Volunteer*. Available at: https://www.witi.com/halloffame/359632/Elizabeth--Feinler-Internet-Pioneer-Active-Volunteer/ (Accessed 30 September 2025
70. Twice.com. (n.d.). *CTA Celebrates 20th Hall of Fame Awards*. Available at: https://www.twice.com/industry/cta-celebrates-20th-cta-hall-of-fame-awards-with-6-inductees (Accessed 30 September 2025).

Speed of Need: Unifying Agile, DevSecOps, MBSE, MOSA, Digital Engineering, and AI for Defense Systems

Robin J. Yeman

Abstract In an era where speed and adaptability are critical to national defense, the U.S. Department of Defense (DoD) must overcome the limitations of fragmented technology adoption and legacy bureaucratic structures. Despite investing in powerful tools and methodologies from Agile and DevSecOps to Artificial Intelligence (AI) the DoD is unable to harness their full potential due to a persistent sociotechnical divide. This chapter argues that modern defense challenges cannot be solved through technical innovation alone; they require the integration of people, policy, and technology into a unified, adaptive enterprise. Through a detailed analysis of recurring failures in acquisition and integration supported by case studies, this chapter exposes how siloed processes, misaligned incentives, and risk-averse cultures continue to delay delivery, orphan innovations, and undermine mission effectiveness. It presents a holistic digital engineering framework that links modular design, continuous delivery, and AI-augmented workflows with organizational transformation, adaptive governance, and empowered human–machine teaming. By aligning sociotechnical elements across the enterprise, the DoD can move from static program execution to dynamic capability evolution delivering trusted, mission-ready systems at the speed of relevance.

1 Introduction

The briefing room at Crystal City buzzed with cautious optimism. Lieutenant Colonel advanced her PowerPoint slide, revealing impressive metrics: 24 software releases in 18 months, automated testing covering 94% of the codebase, and a DevSecOps pipeline that deployed new capabilities every two weeks. The Tactical Operations System (TOS) development team had done everything right. They'd adopted Agile, embedded cybersecurity from day one, used model-based systems engineering to maintain traceability, and built their architecture using open standards.

R. J. Yeman (✉)
Homeland, Leidos, Winter Park, FL, United States
e-mail: robin.yeman@leidos.com

A. Weissman (ed.), *Women in Advanced Computing and Software for Defense*, Women in Engineering and Science, https://doi.org/10.1007/978-3-032-17466-6_3

"We're ready for operational deployment," the Colonel concluded. "The warfighters in the field are asking for these features. We can deliver them next month."

The silence that followed was not the kind that precedes applause.

The program executive officer leaned back, arms crossed. "Your Milestone C documentation isn't complete. We're six months from the next Defense Acquisition Board review." The test director added, "My team hasn't validated these capabilities against the approved test plan. That's another eight months minimum." The contracting officer chimed in: "The contract doesn't have CLINs[1] for incremental delivery. We'd need to modify the agreement, which requires a legal review and approval chain that takes ninety to 120 days".

But the most devastating comment came from the systems integrator in the corner: "Our hardware team is still on the previous baseline. Your software is incompatible with the current fielded configuration. Integration won't happen until the next platform upgrade cycle 18 months out."

Eighteen months. The tactical environment that had driven those 24 releases would evolve a dozen times over. The warfighters who had collaborated with her team would deploy with obsolete capabilities while the new ones sat on a government server, technically perfect and operationally irrelevant.

Six weeks later, a commercial tech company released a similar capability to civilian first responders. Their time from concept to deployment: four months.

Widening Delivery Timelines

The average Major Defense Acquisition Program (MDAP) now requires almost 12 years from program start to initial capability an average that includes MDAPs that began as Middle Tier of Acquisition programs representing an 18-month increase from the previous year [1]. In one assessment period, combined costs increased by $49.3 billion across 30 major programs [2]. Over half of 26 MDAPs assessed that had yet to deliver capability reported new delays driven by supplier disruptions, software development delays, and quality control deficiencies integration and organizational breakdowns rather than technical failures [3].

Even programs using the MTA pathway designed for rapid delivery face this sociotechnical divide. Most MTA acquisition strategies do not outline how programs plan to leverage leading practices to facilitate speed, with some expecting 5 years for rapid prototyping followed by another 5 + years of development [1].

F-35: Integration Challenges Despite Technical Achievement

The F-35 program demonstrates how modern practices collide with legacy structures. The Block 4 modernization effort exceeds original cost estimates by over $6 billion and trails schedule by at least 5 years [4]. In 2024, contractors delivered all 110 aircraft late by an average of 238 days, a significant increase from 61 days in 2023[4]. Technology Refresh 3 (TR-3), a $1.8-billion hardware and software upgrade

[1] Contract Line Item Number (CLIN) is a unique identifier assigned to specific items or services within a contract.

critical to modernization, has experienced setbacks from supply chain disruptions and software integration challenges [5].

DoD completed the final phase of original F-35 development in March 2024 over a decade delayed and $250 billion above original acquisition estimates[6]. These delays stem not from technical inability but from systemic integration challenges across organizational boundaries.

Organizational Incentive Misalignment

The Defense Innovation Board's 2024 study identified "a persistent incentive to maintain the status quo throughout the DoD," [7] noting that leaders are not providing top-down support to innovators and highlighting disconnects between industry, DoD, and the warfighter [8]. According to Admiral Mike Mullen, "The [Defense Department] is very risk averse, and the requirement is to move very rapidly, which is going to require a significant increase in the appetite for risk" [9].

Kessel Run: Demonstrated Capability Within Organizational Constraints

The Air Force's Kessel Run software factory demonstrates both modern practices' potential and scaling barriers. Kessel Run reduced initial software procurement and deployment from the typical three years to three months [10], scaling from 20 personnel to over 700 while delivering capability updates with an eight-hour deployment cycle [11]. In August 2021, the organization supported evacuation of 123,000+ people from Afghanistan through real-time software updates [12].

However, organizational constraints persist. The eight-hour deployment cycle includes significant time burning CDs and manually rescanning artifacts across network air gaps processes dictated by organizational security policies rather than technical limitations [12].

Systemic Pattern

Defense organizations successfully adopt modern technical practices yet deploy them into social systems designed for sequential, document-heavy development. The quantified results include

- 12-year average delivery timelines despite "rapid" pathways [1],
- $49.3 billion cost growth across 30 programs in one year [2],
- 238-day average aircraft delivery delays [4],
- 5-year delays in software modernization efforts [4], and
- MTA programs designed for 2–5 year delivery requiring 10+ years [1].

Government Accountability Office (GAO) concludes: "While the Department of Defense plans to invest more than $2 trillion to develop and acquire its costliest weapon programs, it continues to struggle to deliver new technologies quickly, even while faced with constantly evolving threats" [1]. These are not isolated failures but represent systemic dysfunction. Technological modernization without organizational transformation cannot deliver capabilities at the speed of need.

1.1 Background

The United States Department of Defense (DoD), the most powerful military organization in history, is losing a critical race against time. This conclusion is based on a methodical analysis of the Government Accountability Office's (GAO) annual weapon systems assessments from 2016 to 2024. This body of work reveals a failure not just of process, but of a complex sociotechnical system, the intricate, interdependent web of people, culture, policies, and technologies that must work in concert to deliver capabilities. Despite a planned investment of over $2 trillion, the system is producing declining results. This isn't a temporary setback; it is a systemic failure that threatens America's national security.

Successive waves of reform have proven to be little more than incremental adjustments because they consistently fail to address the acquisition challenge wholistically. The analysis of the 2016–2024 GAO assessments reveals a litany of recurring failures that persist because our solutions are piecemeal [1, 3, 13–20]. We see major programs pushed forward with immature technology, while our people and policies are not equipped to manage the resulting uncertainty. We see a persistent struggle to adopt modern software practices (a technical challenge) because of a risk-averse culture (a social challenge). By treating these as isolated problems, we are merely patching symptoms. The core issue is that you cannot fix the technical elements if the social elements culture, incentives, and expertise remain unchanged.

The root causes are deeply embedded in the powerful interplay between the social and technical dimensions of the system [21]. This is the critical insight that past reforms have missed. They have focused on adjusting a policy or implementing a new framework the technical components without fundamentally altering the human and organizational behaviors that govern them. The system's legacy processes (technical) are protected by a powerful cultural inertia and misaligned incentives (social). This dynamic ensures that even promising reforms are ultimately diluted and defeated by the very system they are meant to change. It is a classic sociotechnical trap: the social system rejects changes to the technical system.

Therefore, the path forward cannot be another narrow, incremental reform. It requires a holistic strategy that finally treats acquisition as the complex system it is. The solution must simultaneously address all core elements, as a failure in one will undermine progress in the others. This is why a transformation built on three integrated pillars is essential:

1. **People (The Social)** Investing in and empowering a twenty-first-century workforce with the skills, autonomy, and incentives to drive change.
2. **Policy (The Social)** Rewriting acquisition rules to be agile and modular, fostering a culture of innovation and managed risk-taking.
3. **Technology (The Technical)** Fully embracing the digital revolution through modern tools and architectures that enable the empowered workforce and flexible policies.

The evidence presented in nearly a decade of GAO reporting is clear. We are not just managing a collection of problems; we are confronting a single, dysfunctional sociotechnical system. Continuing to apply superficial fixes is an abdication of responsibility. A comprehensive transformation that addresses people, policy, and technology in unison is not merely an opportunity for efficiency; it is a national security imperative.

1.2 Context

The Department of Defense (DoD) functions as a sociotechnical system which is defined as a system incorporating humans and technology [22] and is typically used to describe an organizational framework in which social elements (people, culture, policies, and practices) and technical components (tools, technologies, infrastructure, and systems) are deeply interdependent and co-evolving. In a sociotechnical system, success is not achieved by optimizing technology alone but by ensuring that the social and technical subsystems are designed to work together in harmony. This perspective emphasizes that every technical advancement must consider its impact on human operators, organizational structure, and institutional policy and vice versa.

The DoD exemplifies this definition. It is not merely a collection of weapons systems, data networks, and software platforms; it is an enterprise where people from operators and analysts to senior leaders' interface continuously with complex technologies and are governed by layered policies and doctrines. From battlefield operations to logistics and acquisition, outcomes are shaped by the interaction between human judgment, regulatory constraints, and digital systems. Policies influence how technologies are used; culture determines how decisions are made; and workforce readiness affects how innovation is adopted.

In this context, the concept of the **speed of need** becomes a transformative mandate. It refers to the requirement to deliver mission capabilities at a tempo that aligns with evolving operational demands. Crucially, this mandate does not challenge technology alone it pressures the entire enterprise to adapt. Legacy acquisition timelines, rigid policy frameworks, siloed data systems, and hierarchical decision-making processes can all become obstacles to delivering at the speed of need. Therefore, meeting this challenge requires a systemic shift: faster integration of new technologies, more agile policymaking, empowered cross-functional teams, and cultural willingness to embrace iterative learning and continuous improvement.

Understanding the DoD as a sociotechnical system illustrated in Fig. 1 reveals that accelerating capability delivery is not just a matter of developing faster software or smarter hardware it demands reimagining the enterprise itself. The speed of need is thus not just a technical goal, but a call to transform how the entire defense ecosystem operates.

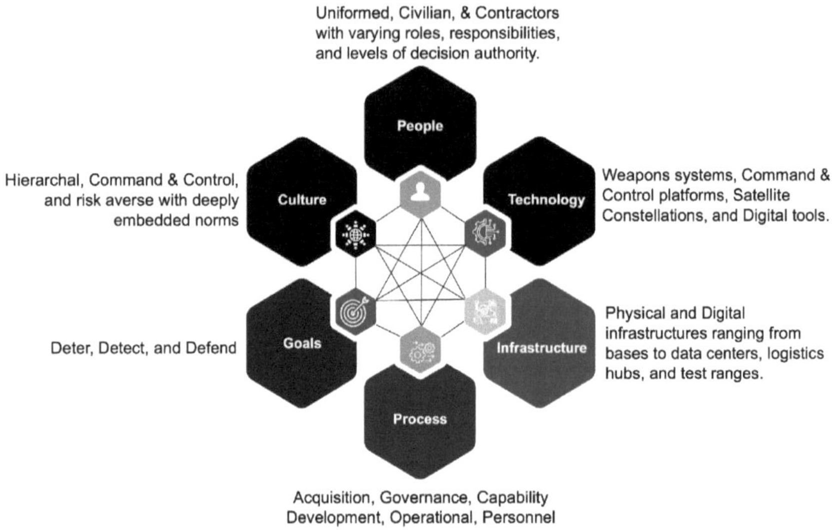

Fig. 1 Department of defense sociotechnical system

1.3 Problem

The Department of Defense (DoD) faces a **two-tiered problem** illustrated in Fig. 2 that severely limits its ability to modernize. The **first tier** involves the fragmented adoption of modern tools and technologies such as Agile, DevSecOps, Model-Based Systems Engineering (MBSE), Modular Open Systems Approach (MOSA), Digital Twins, and Artificial Intelligence (AI). These powerful enablers are often implemented in silos, without coordination across programs, lifecycle phases, or organizational boundaries. As a result, toolchains are disconnected, data is not interoperable, and critical handoffs between functions such as from design to test or development to sustainment become friction points.

The **second tier** of the problem lies deeper: even when tools are adopted, they are not integrated into a coherent sociotechnical system, one that aligns people, processes, culture, governance, and infrastructure with the technology. This leads to misaligned incentives, outdated policies, organizational resistance, and a lack of shared understanding across stakeholders.

These two tiers compound one another: disconnected human systems render fragmented tools ineffective, and unintegrated technologies undermine attempts at organizational change. Without addressing both tiers in tandem, digital transformation efforts will continue to stall, and mission-critical warfighter needs will go unmet.

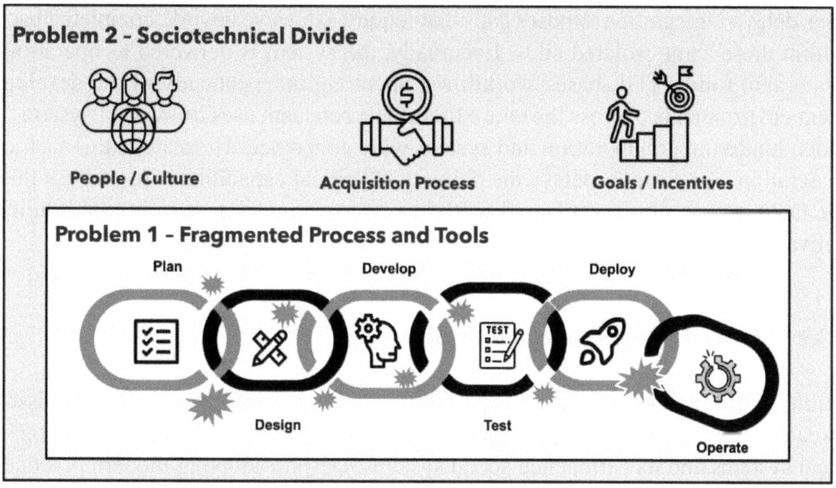

Fig. 2 Two-tiered problem

1.3.1 Fragmented Tooling and Methods

While DoD is adopting transformative modern technologies including Lean, Agile, DevSecOps, MOSA, Model-Based Systems Engineering (MBSE), Digital Twins, and AI to modernize the delivery of warfighting capabilities [23]. However, these powerful practices are often adopted in isolation, creating a fragmented technical landscape where teams work in functional silos. This fragmentation leads to breakdowns at critical transition points, ultimately hindering progress.

This problem unfolds across the development lifecycle. For example, systems engineers may use MBSE and a Modular Open Systems Approach (MOSA) to define architectures, but these artifacts are then transferred to software and hardware teams that lack common tooling or a unified development environment. Software teams may follow Lean and Agile practices, but their output moves to DevSecOps pipelines where security is treated as an afterthought. Meanwhile, hardware teams engage in rapid prototyping, but their development cadence often diverges from the software teams leading to delays.

The emergence of Generative and Agentic AI introduces a dangerous accelerant into this already fractured environment [24]. These AI tools can dramatically speed up work within each silo, generating code, system models, and test cases at an unprecedented rate. Instead of solving the integration problem, however, this speed risks magnifying it. The outputs from one team, now produced faster and in greater volume, will only collide more forcefully with the bottlenecks and misalignments at the next handoff. This creates the conditions for larger-scale integration failures, making the existing friction points even more severe.

When hardware and software components finally converge, they are passed to test teams often too late to meaningfully influence architecture or ensure quality.

This delayed integration exposes gaps that require extensive rework, completed back within those same isolated silos. Eventually, the system is delivered to operations teams that follow ITIL-based workflows, operating independently of the development environments. Across the entire lifecycle, each team uses its own ecosystem of tools, hindering collaboration and system-wide coherence. Ultimately, this lack of integration significantly delays the delivery of critical capabilities. To address this, the DoD must unify its tools and workflows into an integrated, end-to-end digital delivery pipeline before the speed of AI overwhelms its processes entirely.

1.3.2 Divided Sociotechnical System

While tool fragmentation slows capability delivery, a deeper, more systemic issue threatens progress which are the deep disconnects between the DoD's modern technical systems and its entrenched social systems. Despite adopting modern practices such as Agile, MOSA, MBSE, and AI, these efforts are constrained by a risk-averse culture, rigid hierarchies, and dated acquisition models that are fundamentally at odds with agility. This divide [25] results in misaligned incentives, long lead times, and institutional friction that prevent even the most promising innovations from scaling. The root cause is not a failure of technology, but a failure to align people, processes, and policy around a unified operational approach.

This divide is most evident in the language problem is the absence of a shared lexicon across disciplines forcing people to talk past each other. A lack of interaction among different teams leads to fragmented communication [26]. For example, Program Management emphasizes milestone tracking and earned value metrics; Systems Engineers model architectures using MBSE; Software Engineers speak in user stories and Agile sprints; DevOps teams focus on automation and release velocity; and Cybersecurity works from frameworks like the Risk Management Framework (RMF). Compounding the issue, acquisition personnel are bound to linear, document-heavy procurement cycles that cannot accommodate changing needs. The end users such as warfighters often remain distant from the development process, resulting in solutions that may be technically sound but can be operationally misaligned.

This lack of a unified language and shared worldview results in siloed priorities, fragmented toolchains [27], and delayed integration. Bridging this divide requires more than inserting modern tools into legacy organizational structures. Instead, it demands a deliberate effort to redesign the social system to support the technical one by.

- fostering cross-functional literacy among teams,
- empowering decentralized, data-driven decision-making,
- modernizing contracting and acquisition mechanisms for agility, and
- ensuring that end-user feedback continuously shapes development.

Only then can the DoD unlock the full potential of its digital transformation and deliver mission-ready capabilities at the speed of relevance.

1.4 Opportunity

To outpace our adversaries in an era defined by speed, complexity, and relentless competition, the Department of Defense (DoD) cannot simply procure the latest technologies or adopt modern development methods it must fundamentally reimagine its entire approach to capability delivery [28]. The current strategy of layering contemporary practices such as Agile, DevSecOps, MBSE, MOSA, and AI on top of legacy organizational structures and processes has reached its limits. What is needed now is a dual integration strategy beginning by integrating fragmented technical practices into a cohesive, end-to-end digital delivery engine and then to embed that engine within a reformed social system capable of sustaining continuous, scalable, and rapid delivery.

The first layer of integration requires collapsing the silos between engineering, software, hardware, cybersecurity, testing, deployment, and operations through shared models, interoperable toolchains, and synchronized cadences [29]. The risk in system development exists in the liminal areas, where we cross boundaries.

This means linking MBSE and MOSA to Agile backlogs, ensuring DevSecOps pipelines are secure by design from the start, aligning hardware development with software sprints, shifting test and user feedback far left. The second and more difficult layer requires restructuring the cultural, contractual, and organizational context in which that engine operates. Acquisition models must evolve to support iterative delivery and modular integration. Contracts must be reshaped to incentivize outcomes over outputs. Functional teams must develop cross-disciplinary fluency. Data driven decision-making must decentralize to empower those closest to the problem.

Only by integrating the technical and the social system can the DoD deliver capabilities at the speed of need with resilience, agility, and strategic advantage. This is not a transformation, it is a reinvention of how defense capabilities are conceived, built, fielded, and evolved in a dynamic threat environment.

2 Fractured Kill Chain

2.1 Fragmented Technologies

2.1.1 Introduce the Technologies

Lean

Originating from the Toyota Production System in the 1940s and formally articulated in the 1990s, Lean focuses on eliminating waste, optimizing value streams, and improving flow across processes. It emphasizes continuous improvement (Kaizen) and customer value as central tenets [30]. In the defense sector, Lean is used to streamline manufacturing lines, reduce cycle times, and eliminate non-value-added steps in acquisition or logistics. For example, the U.S. Army has applied Lean principles in depot maintenance operations to cut turnaround times and reduce parts inventory [31].

Agile

Agile emerged in 2001 with the publication of the Agile Manifesto, which emphasized iterative development, customer collaboration, and team autonomy [32]. Agile is primarily used in software development but has since expanded into systems engineering, program management, and even hardware development. Its strength lies in delivering small, functional increments rapidly, incorporating feedback, and adapting to change. Many teams leverage Jira, a product backlog planning tool to manage work. An example is the U.S. Air Force's Kessel Run program, which uses Agile to deliver mission-critical software updates to warfighters within days rather than months [33].

DevSecOps

DevSecOps evolved in the 2010s by integrating development (Dev), security (Sec), and operations (Ops) into a continuous delivery pipeline. It emphasizes automation, continuous integration/continuous delivery (CI/CD), and built-in security from the beginning of the development process [34]. DevSecOps is especially relevant in defense environments where security must be assured at every stage of system deployment. For instance, the U.S. Department of Defense's Platform One uses DevSecOps practices to enable secure, continuous delivery of containerized software across classified and unclassified environments [35].

Model-Based Systems Engineering (MBSE)

Introduced in the early 2000s, MBSE is a shift from document-centric to model-centric systems engineering, enabling traceability, simulation, and early validation of complex systems. It uses tools like SysML to represent requirements, architecture, and behavior in a formal, executable manner [36]. MBSE is widely used in aerospace and defense NASA, for example, employs MBSE to model spacecraft systems and reduce integration risk during space missions [37].

Modular Open Systems Approach (MOSA)

MOSA is a Department of Defense initiative formalized in the 2010s by DoD but conceptually rooted in modular design and open interface standards. It promotes interoperability, component reuse, and vendor independence. MOSA is used to enable rapid upgrades, integration of commercial off-the-shelf (COTS) components, and lifecycle cost savings [38]. A leading example is the Navy's Future Airborne Capability Environment (FACE), which applies MOSA principles to avionics systems to ensure interoperability and upgradeability across platforms [39].

Shift Left

Shift Left testing emerged as a software engineering practice advocating for early testing, validation, and security in the development cycle. The goal is to catch issues before they become costly downstream failures [40]. It's commonly used in Agile and DevSecOps environments to integrate static analysis, unit testing, and vulnerability scanning from the start. In defense programs, Shift Left is used to embed cybersecurity assessments during architecture design, such as in the Missile Defense Agency's software pipeline practices [41].

Rapid Prototyping

Rapid prototyping originated in the 1980s with the rise of computer-aided design (CAD) and computer numerical control (CNC) tools. It enables quick fabrication of physical prototypes for concept validation and iteration [42]. It is widely used in defense to accelerate hardware development cycles. For instance, DARPA employs rapid prototyping in programs like Gremlins and Blackjack to quickly evaluate aerospace component designs and iterate based on test feedback [43].

3D Printing

Also known as additive manufacturing, 3D printing was developed in the 1980s, with major industrial adoption beginning in the 2010s. It enables layer-by-layer construction of physical parts, often from CAD models, and is used to reduce tooling costs, enable on-demand part production, and support complex geometries [44]. The U.S. Navy uses 3D printing to produce spare parts aboard ships, improving mission readiness and reducing the need for extensive onboard inventories [45].

Digital Twins

The concept of Digital Twins was first articulated by NASA in the early 2000s to support real-time simulation and monitoring of spacecraft systems. A digital twin is a virtual representation of a physical system that mirrors its behavior, performance, and condition through sensor data [46]. Digital twins are now used across defense and industry to support predictive maintenance, operational planning, and testing. For example, the Air Force uses digital twins to simulate the lifecycle performance of fighter jet components, optimizing sustainment strategies [47].

Artificial Intelligence (AI)

Artificial Intelligence (AI) originated in the 1950s as an academic field focused on creating machines capable of performing tasks that typically require human intelligence, such as reasoning, learning, and decision-making [48]. Over time, AI has evolved to include a wide range of capabilities, including supervised learning, computer vision, and natural language processing. In defense, AI is broadly used for pattern recognition, data analysis, and autonomous decision support. A notable example is Project Maven, where the DoD applies AI to analyze full-motion video from drones, rapidly identifying objects of interest to reduce cognitive load and accelerate targeting decisions [49].

Generative AI

Generative AI, a subfield of AI that gained widespread attention in the 1960s, focuses on creating new content text, images, code, or data using models like generative adversarial networks (GANs) and transformer-based architectures [50]. Generative AI is used in defense for code generation, training simulation content, mission planning, and threat scenario modeling. For instance, the U.S. Air Force and DARPA have explored using generative models to automatically generate synthetic training data to augment real-world datasets for computer vision and sensor systems [51].

Agentic AI

Agentic AI represents the next evolution, where AI systems act with goal-oriented autonomy, proactively planning, executing tasks, and making decisions within defined constraints and intent. This paradigm had its roots in 1960 but has become popular in the 2020s as AI systems became capable of orchestrating workflows, tools, and reasoning processes independently [52]. Agentic AI is especially promising for complex mission environments requiring adaptive behavior, multi-step reasoning, and human-AI teaming. A defense example is the development of autonomous mission planning agents that can interpret commander intent, generate viable plans, coordinate resources, and adjust in real-time based on battlefield conditions augmenting human decision-makers with scalable, responsive AI partners [53].

2.1.2 Siloed Methodologies Clash

Despite widespread adoption of transformative technologies and Modern practices such as Lean, Agile, DevSecOps, MBSE, MOSA, Shift-Left Testing, Digital Twins, and 3D Printing the Department of Defense (DoD) continues to face systemic dysfunction due to isolated implementation within siloed organizations. Each technology or process is powerful, but when applied in a fragmented enterprise without shared language, governance, or coordination, the result is friction instead of acceleration. Modern capabilities clash with legacy processes and outdated oversight structures, eroding their intended value.

The Agile-Waterfall Collision

A team may deliver functional capabilities weekly, but within a Waterfall acquisition program, those deliveries are not consumed or validated for months or even years due to the formal Systems Engineering Technical Reviews (SETRs) associated with the phase gates. These reviews are tightly coupled with how we validate safety, security, and compliance for systems that contain hardware and software [54]. Yeman describes in the application of Agile to large-scale, safety–critical, cyber-physical systems how to move to a continuous approach of validating security and regulatory compliance using the defined continuous assurance plug-in [55]. Currently the delays for capabilities to sit idle waiting for feedback.

The DevSecOps-ATO Mismatch

DevSecOps pipelines may be automated and secure-by-design, yet they are still forced to package capabilities and paperwork in EMASS for static Authority to Operate (ATO) review which can take a year or more completely misaligned with the pipeline's rapid delivery cadence. "ATO is defined within AFI 17-101 as: The official management decision given by a senior organizational official to authorize operation of an information system and to explicitly accept the risk to organizational operations (including mission, functions, image, or reputation), organizational assets, individuals, other organizations, and the Nation based on the implementation of an agreed-upon set of security controls. (Secretary of the Air Force Chief Information [SAF/CN], 2024)" [56].

The Lean versus Bureaucracy Conflict

Lean practices, aimed at eliminating waste and optimizing flow, often clash with bureaucratic oversight that mandates layers of reporting, approvals, and redundant checkpoints [57]. A team using Lean value stream mapping may identify dozens of non-value-added steps, only to find their progress halted by rigid, compliance-driven gate reviews.

The Disconnected Digital Model (MBSE & MOSA)

Model-Based Systems Engineering (MBSE) tools may be used to define system architecture in models like SysML, but if other teams do not contribute to the model, it becomes a disconnected artifact rather than the living, shared source of truth it was intended to be. Worse, if MBSE artifacts are not aligned with Agile development tools or test automation, traceability breaks down, and rework increases [58]. One of the problems described in Caroll and Malins research is the lack of skilled engineers resulting in poor implementation [59]. Despite Modular Open Systems Approach (MOSA) being required by legislation [60], there are still several challenges in the implementation. Some in the defense community use MOSA as a checklist compliance activity. Designing Modular platforms has proven challenges, and many efforts have failed to meet expectations for cost and quality. When MOSA principles are adopted in isolation without shared interface governance or contract language that supports modularity programs still end up with nullifying the benefits of modular design.

The Static Digital Twin

Digital Twins are digital representations of physical object, person, or process contextualized into its environment [61]. Digital twins go beyond modeling by incorporating real-time data from the physical world. A digital twin is a living model that continuously updates in parallel with its physical counterpart, providing a real-time mirror of the actual system, asset, or process. Digital twins integrate IoT sensors, machine learning, and analytics to simulate the current state of their physical twins accurately and predict future states [62]. The benefit of Digital twin is the ability to safely validate multiple approaches when updating physical systems. The challenges are there is limited adoption across the DoD due to cost, complexity, and potential cyber security impacts [63]. In addition most programs do not invest in building until very late in the lifecycle.

The Legacy Block on 3D Printing

3D Printing can create physical objects from geometric representation typically in a computer aided design (CAD) file through the successive addition of materials [64]. 3D printing is playing a significant role in the defense efforts of both the US Department of Defense (DoD) and Ukraine, enabling rapid production of parts and equipment, particularly in the context of Ukraine's war with Russia. While 3D Printing enables rapid hardware prototyping and part production, when we are not in an emergency conflict it hits a wall of legacy supply chain, quality assurance, and certification protocols that are not adapted for additive manufacturing workflows leading to delays, re-verification, and wasted materials.

Shift-Left Testing, Applied Too Late

Shift-Left began with shifting test early in development to catch issues sooner, and later the Airforce expanded the definition to shift left safety, security, airworthiness, and other regulatory constraints described in Dr. Will Ropers Paper, "There is no spoon" [65]. The benefit of shifting left is that we are beginning with the end in mind and building safety, security, and overall quality into the process while reducing lead time to value delivery [66]. However, this is frequently met with resistance due misunderstanding of the value, lack of trust, low tolerance for change, or simply cognitive overload without adequate time to learn [67]. Other problems that include test teams are brought in too late or operate in separate toolchains from developers. In these cases, test insights fail to inform design decisions, and defects are discovered far downstream negating the intended time and cost savings.

Together, these examples highlight a critical truth: the DoD is adopting modern tools built for speed, agility, and modularity, but deploying them in systems still governed by industrial-era processes, stove-piped responsibilities, and rigid oversight structures. Until technical capabilities are fused into integrated delivery engines, and those engines are embedded within a reformed socio-organizational structure, the full promise of these innovations will remain unrealized. Instead of exponential value, we get linear gains trapped in circular bureaucracy. The rise of Generative and Agentic AI offers an unprecedented opportunity for the Department of Defense (DoD) to accelerate capability delivery, enhance decision-making, and reduce cognitive and manual burdens across the enterprise. Generative AI can automatically create code, generate synthetic data, and simulate operational scenarios in minutes. Agentic AI takes this further, enabling systems that reason over goals, plan actions, and autonomously orchestrate complex workflows. These tools are designed to function within tightly integrated digital ecosystems ones that allow for real-time interaction and continuous learning. Yet, the DoD's fragmented landscape of tools, data, and workflows is the antithesis of the integrated ecosystem these AI systems require to thrive.

The Productivity Paradox: Speed in a Vacuum [68]

When AI tools are introduced into siloed development environments, their productivity gains are often nullified. A Generative AI might boost a developer's throughput by 50%, but that value evaporates when the software enters a months-long manual accreditation process [69]. Agentic AI has shown to improve productivity completing tasks from data analysis to complex planning by 45% [70]. Agentic AI could implement multiple tasks across functional areas, but without integration of the functional areas the benefits will be lost in the silos. Agentic AI is a capability that could swiftly help the DoD manage changing battlespace conditions and help solve large-scale, complex problems independently. This differs from current popular large language models dependent on individual prompts to perform a simple, specific task [71]. The Russia-Ukraine war has offered a glimpse of the value of AI in modern warfare and its impact on military operations and tactics. Earlier this year, Time reported that Palantir Technologies AI software was responsible for most of the targeting in Ukraine [72].

The Crisis of Trust: Fragmentation and Risk

This fragmentation also undermines the trust and reliability of AI-generated outputs. Without unified data governance and interoperable toolchains, Generative and Agentic AI are forced to reason over incomplete, stale, or inconsistent data inputs, leading to hallucinated outputs, incorrect recommendations, or decision delays. In the high-stakes context of national defense, this is an unacceptable risk. For example, an AI system that generates predictive maintenance plans based on sensor data will fail if logistics, test, and sustainment teams operate in separate tools without synchronized models. The AI may offer the right answer, but the system around it is incapable of acting on it quickly or safely.

A Mandate for Integration

To harness the true potential of Generative and Agentic AI, the DoD must break down the silos across the value stream between acquisition, systems engineering, software engineering, hardware engineering, test engineering, and operations. Ineffective integration of process and tools leads to fragmented operations that hinder the seamless flow of information, making it challenging to pivot or innovate in response to external pressures [73]. We have an opportunity to leverage Artificial Intelligence (AI) to break down the functional area silos by democratizing access to knowledge across the organization [74]. This improvement could dramatically improve the flow of value delivered into the hands of the warfighter. If we implement the power of AI within the silos, we will minimize the improvements seen because we have not removed the bottlenecks in the organization. AI systems require not just data access but system-wide context across the entire delivery value stream from concept to operations. AI isn't just another tool it's a capability multiplier that demands cohesive integration of tools, people, and processes. Without this integration, the DoD will squander its AI investments by applying next-generation intelligence to a legacy machine incapable of keeping pace with the speed, scale, and complexity of modern warfare.

2.1.3 Consequences

Despite the DoD's adoption of modern tools and ways of working the fragmented adoption undermine their effectiveness. When these technologies and methods are introduced in isolation, disconnected from one another and from the overarching acquisition and operational ecosystem, they generate friction instead of flow. The consequences are severe: brittle systems, orphaned innovations, program delays, cost overruns, and reduced warfighter effectiveness. In a rapidly evolving global threat environment, these outcomes are not just inefficient they are strategically dangerous.

Brittle Systems and Delayed Capability Delivery

Poor integration across tools and workflows leads to brittle systems that are fragile, inflexible, and unable to adapt to change [73]. These systems often appear robust during planning but collapse under the weight of real-world complexity, late-stage integration, and evolving requirements. The F-35 Joint Strike Fighter program is one such example. Initially envisioned as a common, modular platform, the program suffered years of delays and spiraling costs due, in part, to integration challenges across hardware, software, test, and logistics systems [76, 77]. Despite using advanced technologies like model-based engineering, the program was plagued by incompatible toolchains and disconnected teams, resulting in poor subsystem integration and years-long lag in delivering promised capabilities.

The Valley of Death

Innovations introduced in isolated environments without full lifecycle integration frequently become orphaned technologies capabilities that work in isolation but are never fielded or scaled, this is referred to as the "Valley of Death". According to Deputy Defense Secretary Kathleen Hicks, transition of innovative technologies is one of the militaries biggest problems [78]. This problem is even seen in critical technologies such as Generative and Agentic AI. Colin Crosby in the Marine Corp stated that "It's not that people are afraid of the technology, they just don't understand how the technology can work for them." The deputy chief data and AI officer of the coast guard stated that we need demystify AI for their personnel [79].

Lack of End-To-End Traceability and Decision Latency

Fragmentation also eliminates traceability the ability to connect requirements to design, test, fielding, and operations. Without traceability, decisions are made in isolation, feedback loops are broken, and accountability is diffused. The Navy's Littoral Combat Ship (LCS) program is illustrative: envisioned to be fast, modular, and upgradable, the program was bogged down by integration issues between mission packages and core systems [80]. The result was years of delays and reduced mission capability.

Strategic Risk: Inability to Respond at the Speed of Need

The biggest consequence of poor integration is the DoD's inability to deliver capabilities at the speed of need. The GAO report for 2024 stated "DOD remains alarmingly slow in delivering new and innovative weapon system capabilities" [81]. In modern warfare, where adversaries use iterative technologies and adaptive tactics, delivery cycles measured in years are unacceptable. In 2024 China files 300,510 AI-related patents, in the same period the United States files only 67,773 showing that in the Age of AI they are outpacing the US 5 to 1 [80]. Failing to develop, integrate, and scale technology at the pace of our adversaries is measured not only in dollars and delays but in operational readiness, mission impact, and strategic deterrence. Without both intentional and cohesive integration, modern technologies become incompatible parts in a disjointed machine. The only path forward is to unify technical systems, data pipelines, and human workflows into a coherent, adaptive, and digitally integrated delivery engine that can evolve as fast as the threats it is meant to counter.

2.2 *Sociotechnical Divide*

2.2.1 The Barrier Beyond Tools

The investments that the DoD has made in modernizing tools and techniques have not resulted in the expected speed, flexibility, and mission effectiveness that was promised as illustrated in the GAO 2024 Weapons Systems Annual Assessment. The average Major Defense Acquisition Program (MDAP) delivery cycle time has increased from ~8 years to now ~11 years [82]. The reason is not a failure of tools or methods, but a deeper, systemic issue which I refer to as the sociotechnical divide. The divide refers to the misalignment between modern technical practices and the legacy social systems in which they are deployed.

This includes legacy governance, risk-averse cultures, rigid hierarchies, and acquisition frameworks that were not built for the speed of digital and definitely not built for the Age of AI, where 90% of the world's data was generated in the past 2 years [83]. While development teams iterate rapidly, they are constrained by governance models that demand milestone reporting instead of measuring the continuous delivery of value.

DevSecOps pipelines may automate integration and deployment, but still require manual approvals designed for waterfall-based change management. MBSE models may define exquisite architectures, but if downstream teams are not integrated into the modeling environment, the model becomes static rather than executable. In addition, they are only understood by systems engineers, which leaves a lot of the world out.

The structural and cultural mismatches have resulted in a brittle system, where change is difficult, integration is delayed, and innovation is contained within silos. Teams operate with modern tools but are governed by legacy rules, leading to frustration, rework, and disillusionment with transformation efforts. Failing to address the divide causes organizations to misdiagnose the problem, treating transformation as a tooling initiative as opposed to the reality of the problem which is a comprehensive re-architecture of how people, processes, and policies align to deliver value. Until the DoD addresses this systemic disconnect, even the most advanced technologies will fail to deliver their transformational promise.

2.2.2 Babylon with the Absence of a Common Language

One of the most visible and persistent symptoms of the sociotechnical divide is the absence of a shared language across functional teams. This is not merely a matter of different terminologies, but of fundamentally different mental models and measures of success. For example, success means different things depending on one's role. For an individual team member, it means completing a task or user story. For a team, it means increasing velocity or meeting sprint goals. At the organizational level, success means achieving revenue growth or strategic outcomes. For a DoD procurement it means delivering a capability on time and within budget. Finally, for the end-user, success means using the product to effectively accomplish their mission. True value is delivered across the system only when these diverse perspectives are understood and aligned. The problem is exacerbated when we have siloed functions.

When teams are siloed, each focusing solely on their own domain like systems engineering documents, software, or hardware, without a shared, aligned understanding of success, several critical issues can emerge:

- **Misaligned Objectives** Each team may optimize for its own definition of success (e.g., completing documentation, writing code, or delivering hardware), rather than contributing to the overall system value or end-user mission success. This leads to outputs that meet local goals but fail to integrate effectively.
- **Integration and Rework Costs** Siloed work often leads to late-stage integration problems where components don't function cohesively. This results in rework, cost overruns, and schedule delays undermining program-level success.
- **Incomplete Traceability and Poor Validation** When documents, models, and components are developed in isolation, it becomes difficult to maintain end-to-end traceability. This impairs verification, validation, and regulatory compliance, especially in safety–critical systems.
- **Reduced Agility and Slow Feedback Loops** Without cross-functional collaboration, feedback from downstream users or stakeholders arrives too late, limiting the ability to adapt quickly to changes in requirements, technology, or mission priorities.
- **Loss of User-Centered Focus** Teams may meet their specific deliverables without ensuring the product supports the user's operational goals, reducing overall effectiveness and user satisfaction.

Within the DoD, each domain speaks in its own professional dialect. These differences exacerbate siloed thinking and create formidable barriers to collaboration. In Table 1 are a subset of these functional areas.

Table 1 Example functional areas

	Functional domain	Focus area	Language
1	Acquisition	Federal Acquisition Regulation (FAR), focusing on contract deliverables, fixed scopes, and milestone-based funding	FAR, contracting
2	Program management	Focuses on earned value, risk management, and milestone attainment	PMI, Prince2
3	Systems engineering	Requirements, MBSE, prioritizing functional interfaces and traceability	Systems thinking, INCOSE
4	Software engineering	Speaks in building code and measuring success by CI/CD pipeline throughput	Agile
5	Hardware engineering	Prioritizes physical design constraints (SWaP-C), component lead times, and producibility	IEEE, ASME
6	Cybersecurity	Implements the risk management framework (RMF), emphasizing control baselines and achieving authority to operate (ATO)	OWASP, CSA
7	Test and evaluation	Relies on formal verification protocols and validation metrics	ISTQB
8	Operations	Prioritize uptime and system support using ITIL-based frameworks	ITIL, ITSM

While each group's priorities are valid within its domain, the lack of a shared language ensures they inevitably collide rather than align. The result is a tangled web of misaligned priorities, conflicting success criteria, and fractured toolchains. A team delivering valuable software every two weeks may find its work sitting idle, ignored by an acquisition process prepared only to accept capability at formal, annual milestones. MBSE models may define a complete system architecture, but if not integrated with test and operations tooling, they offer little real-world validation. Cybersecurity may bolt on controls late in the lifecycle because they were excluded from design conversations due to different timelines and toolsets.

These language and worldview mismatches lead to duplicate work, rework, and late-stage integration failures that slow programs and frustrate teams. What begins as a semantic disconnect metastasizes into a direct delivery risk. Without a deliberate effort to foster cross-disciplinary understanding and create a unified delivery lexicon, the divide will only deepen especially as the DoD integrates increasingly complex technologies such as Generative AI and Agentic AI, which require trust, transparency, and shared context to be effective.

Important Fictional Vignette An Agile software team supporting a mission-critical communications platform completes a sprint early, delivering a functional module that enables encrypted message routing. They log the release in their CI/CD dashboard, mark the user story complete in Jira, and prepare for a demo. However, the systems engineering team, operating in SysML, does not update the interface control document (ICD) to reflect the change so downstream hardware teams continue building against an outdated model. Meanwhile, the cybersecurity team is unaware of the sprint delivery because it hasn't been submitted as part of the ATO package, which is still in draft. The program manager, expecting formal milestone documentation, is unaware of the release because it doesn't align with the integrated master schedule (IMS). Finally, the contracting officer cannot authorize further work because the functional delivery was not part of the contract line items (CLINs), which are milestone-bound. The result: functioning software sits undeployed, feedback is delayed, and momentum is lost not because the teams aren't working hard, but because they're working in different worlds.

2.2.3 Frozen Middle and Cultural Inertia

One of the most significant barriers to digital transformation in highly regulated organizations such as the DoD lies not at the edges of the organization, but in its center in the middle management layer, often referred to as the frozen middle. In a speech by deputy secretary of defense Kathleen Hicks that our well-intentioned disruptors need to thaw the frozen middle, embrace them, and hold them accountable [84]. While senior leaders advocate for innovation and modernization transformation efforts frequently stall at this intermediate layer. This stall is not a failure of intent but a consequence of a system that incentivizes this group to minimize risk and prioritize control and compliance over the adaptability and speed that modern warfare demands. In 2024 the Defense Innovation Board (DIB) published that the DoD incentive structures are broken, consequently obstructing innovators from adopting new technologies faster [85]. Middle management roles are often governed by legacy performance metrics rooted in historical thinking such as adherence to schedules, completion of documentation, milestone achievement, and risk mitigation through oversight. These structures are fundamentally incompatible with modern engineering paradigms that embrace emergence, iterative development, and decentralized decision-making. This conflict places middle-tier leaders in an impossible position: they are tasked with executing rapid delivery while being measured by rigid, legacy governance frameworks. Unlocking the DoD's potential requires a new strategy, not to bypass the frozen middle, but to thaw it by actively enabling it. This includes aligning middle-management performance metrics with flow-based delivery, training leaders in digital fluency, and granting them the autonomy and trust to operate within modern value streams. Ukraine is an example of how this succeeds, their battlefield has evolved into an innovation lab for obtaining modern defense technologies [86]. According to their armed forces representative, if any innovation has not reached production in less than 6 months, they risk becoming obsolete and orphaned.

2.2.4 Acquisition Process

A critical structural barrier reinforces the sociotechnical divide in the DoD's legacy acquisition and contracting system designed for an era of physical manufacturing.

The DoD's acquisition framework anchored in the Federal Acquisition Regulation (FAR) and Defense Federal Acquisition Regulation Supplement (DFARS) is fundamentally misaligned modern development. One of the key changes is that we live in a software defined hardware enabled world. Software defined is known as programmable hardware that allows for configuration and management of hardware through software [87]. Software defined provides the ultimate flexibility and adaptability in development. While previously hardware and software were developed separately the new paradigm performs integration from onset of design. Software behaves very differently from hardware therefore the current acquisition approach does not handle well. In 2018 a new software acquisition pathway was added to the arsenal of tools offered by the DoD but it was not leveraged frequently. In March of 2025, Defense Secretary Pete Hegseth issued a memo that mandates the use of the pathway. Stating that software is at the core of every weapon system and we need to go faster to remain the most lethal fighting force in the world [88]. This relates back to the fact that weapons are software defined and hardware enabled.

The traditional acquisition approach directly conflicts with modern development approaches. Programs that attempt to operate iteratively must still secure funding through Program Objective Memorandum (POM) cycles that require five-year forecasts, issue Contract Line-Item Numbers (CLINs) tied to rigid deliverables, and pass milestone reviews that do not account for emergent learning. Five-years is a lifetime at the current pace of change both in the geopolitical climate and with the current pace of technology change. The acquisition workforce is overworked and lack time to understand all contract types and modern delivery models. Contracting officers, program integrators, and budget officials often lack exposure to modern technologies [89]. This limits their ability to write, manage, or evaluate contracts that align with modern practices. Even where intent to modernize exists through efforts like the Adaptive Acquisition Framework illustrated in Fig. 3, the application remains inconsistent.

To enable true agility at scale, the DoD must modernize its acquisition ecosystem alongside its technical one. This includes rethinking how contracts are structured, how success is measured, and how acquisition professionals are trained. Contracts must evolve from instruments of control, focused on documentation, to instruments of enablement, focused on delivering outcomes through iterative, modular value [89]. Until this evolution occurs, even the most capable digital programs will remain

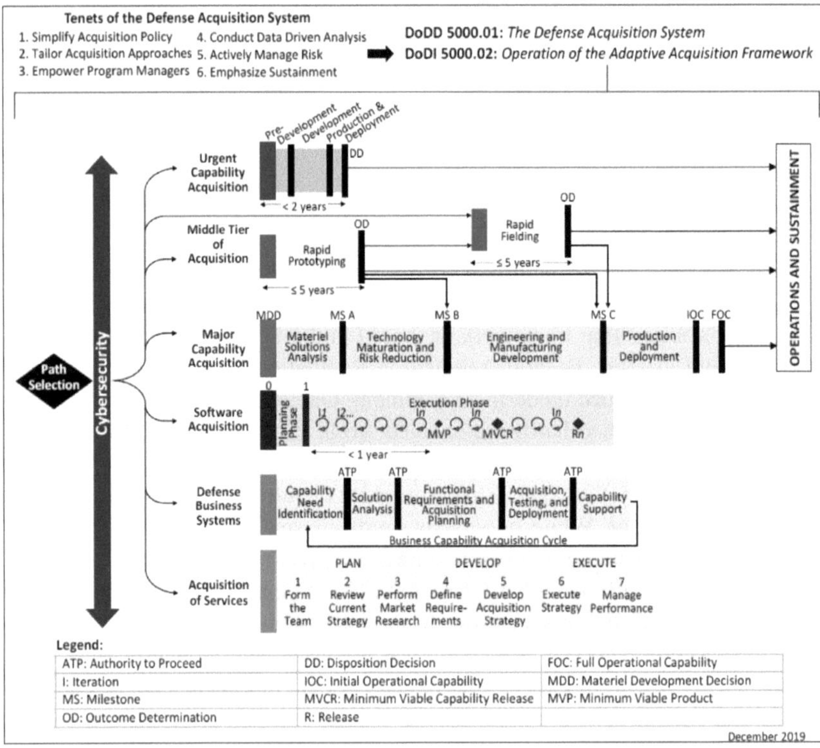

Fig. 3 Adaptive acquisition framework. *Source* U.S. Department of Defense, DoDI 5000.80 (2019) [90]

constrained by outdated policy, governance, and oversight models that cannot support the speed, flexibility, or innovation demanded by today's defense environment.

2.2.5 AI May Magnify the Trust Gap

As DoD accelerates its adoption of advanced Artificial Intelligence, specifically in the areas of Generative and Agentic AI, it confronts a new and deeply human dimension of the sociotechnical divide: the dual challenge of trust and fear [91]. While these technologies promise to revolutionize how the DoD builds systems, makes decisions, and executes operations, their effectiveness is ultimately gated by these same organizational and cultural challenges. In addition, without addressing this human dimension head-on, the DoD risks investing in powerful AI that is ultimately undermined by a rational fear of the unknown, misalignment with existing processes, and deep-seated institutional inertia.

While Generative AI can accelerate the creation of code and documentation, and Agentic AI can autonomously plan and execute workflows, both capabilities collide with an organization unprepared to trust them. This institutional friction often manifests as rational fear and distrust:

- **Cybersecurity teams**, fearing hidden vulnerabilities or compliance breaches, will distrust code they did not personally review line by line.
- **Acquisition officers**, fearing contractual protests or audit findings, will reject AI-generated artifacts that do not map neatly to predefined deliverables.
- **Test and evaluation personnel**, fearing unreliable or irreproducible outcomes, will lack confidence in results they cannot independently verify through traditional means.
- **All domains**, there is an underlying fear that AI will render deep-seated professional expertise obsolete, threatening established roles and career identities.

These breakdowns are not technical; they stem from social systems unprepared to absorb AI's speed, scale, and autonomy.

This trust gap is exacerbated by a deeper, more personal factor, a rational fear of professional obsolescence and loss of control. Many professionals particularly in the acquisition, compliance, and governance communities see AI not as an enabler, but as a threat. Some fear that AI will render their skill sets obsolete [92], while others, rightly, are hesitant to delegate critical decisions to "black box" systems whose logic they cannot inspect or understand [93]. These concerns are valid and, unless addressed through transparent communication and clear ethical boundaries, will reinforce resistance. When trust is absent, AI outputs are dismissed, adoption stalls, and the immense investment in these technologies is squandered.

Compounding this fear is the reality that most current processes are not equipped to interface with AI-native workflows. Generative AI may produce production-ready code, but it cannot bypass outdated manual approval pipelines. Agentic AI may recommend strategic changes in minutes, but organizations accustomed to top-down command structures are unlikely to act without layers of human validation. This friction creates an unacceptable drag on decision-making. Ultimately, trust in AI is not a measure of the algorithm's performance; it is a measure of the organization's readiness to interpret, verify, and act on AI-driven insights.

To close this gap, the DoD must develop human-centered integration strategies for AI. This includes investing in AI literacy across roles, aligning incentives to reward human–AI collaboration, and creating oversight models that combine accountability with adaptability [94]. Addressing the human fears of malfunction, job displacement, and professional irrelevance is just as critical to mission success as addressing algorithmic bias. Without that shift, Generative and Agentic AI will not bridge the divide, they will deepen it.

3 Forging the Solution

3.1 Begin by Integrating the Technical Engine

3.1.1 Ideal Technical Workflow

We begin by integrating the tools and technologies across the product delivery value stream illustrated in Fig. 4.

The DoD faces an imperative to deliver capabilities faster, with greater adaptability and assurance, in a threat landscape that evolves at machine speed. Meeting this challenge requires not just adopting new technologies in isolation but integrating them across the entire value stream. The ideal workflow knits system design, development, and deployment into a seamless digital ecosystem that eliminates silos and enables the continuous delivery of mission-ready capabilities. This workflow blends modularity, model-based design, agility, automation, and operational insight into a cohesive delivery engine.

The Foundation: Modular Architecture and Model-Based Design

At the foundation of the DoD reimagined workflow are Modular Open Systems Approach (MOSA) principles, which allow the ability to replace weapon parts faster [60]. While this may introduce short-term costs at the start it will provide substantial benefits in the future. These architectures are captured and validated using Model-Based Systems Engineering (MBSE). Through executable models often built using SysML and LML MBSE connects system requirements, behaviors, and structural designs in a way that ensures consistency, traceability, and early validation. In

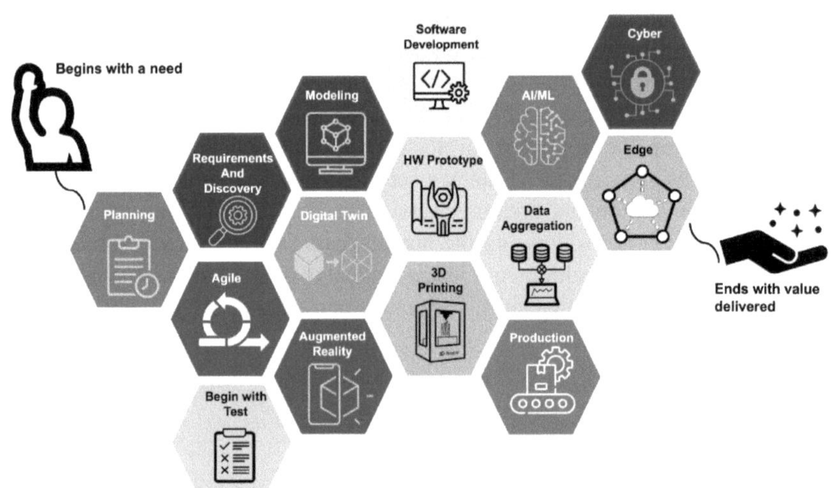

Fig. 4 Integrated value stream

essence, MBSE allows teams to visualize, simulate, and "test" critical decisions before committing to costly hardware or software development. MBSE provides both early-term efficiency in terms as well as long term benefits such as optimizing system design, reducing time, costs, and reducing technological risks, and enhancing the efficiency of business systems engineering through lifecycle management and quality management [95].

The Execution Engine: Agile Delivery

These architectures and models feed directly into Agile backlogs, where user stories and features are continuously prioritized based on mission needs and evolving requirements [95].

MBSE supports Agile implementations by providing traceability typically needed for compliance in safety critical systems. Agile delivery is organized around rapid iterations that break down large system functions into small, testable increments. Instead of long development phases and late-stage integration, each sprint delivers value and learning that informs the next. This allows teams to adapt, align with real-time operational feedback, and deliver partial capability without waiting for full system completion.

The Automation Pipeline: DevSecOps at Scale

Workflows from the backlog into an automated DevSecOps pipeline that ensures speed, security, and quality. Continuous Integration and Continuous Deployment (CI/CD) pipelines build, test, and deploy software automatically, embedding security (shift left) as early as code commit. This automation extends to hardware development as well, using model-linked build plans, additive manufacturing (3D printing), and rapid prototyping frameworks that keep pace with software delivery cadences. Cybersecurity, safety, and compliance are not external reviews; they are embedded within the pipeline through static analysis, dynamic testing, and validation against control frameworks at every increment.

Continuous Assurance: Integrated Test and Evaluation

Lastly, Test and Evaluation (T&E) ceases to be a final, isolated gate; instead, it becomes a continuous process that begins on day one. Integrated test plans are derived from system models and requirements, and automated tests are triggered in the DevSecOps pipeline. This ensures that quality, compliance, and system assurance evolve with the system itself, rather than being retrofitted at the end. Embedding test early also facilitates regulatory approval and accreditation, which can proceed incrementally rather than as a singular final milestone.

The Virtual Proving Ground: Digital Twins

To bridge the physical and digital domains, the workflow incorporates a Digital Twin environment a real-time, sensor-fed virtual representation of the system under development. Digital Twins enable teams to stress-test behavior, validate assumptions, and observe performance under realistic operational conditions long before a physical system is fielded. They are critical for predictive maintenance, operator training, and system tuning. Because they are fed by live data and linked to system models, they serve as a dynamic thread connecting design intent with field reality.

Seamless Integrated System Delivery

Finally, the workflow culminates in Integrated System Delivery, where hardware and software converge seamlessly. This convergence is not left to a late-stage integration effort; it is managed continuously through aligned development rhythms, shared data environments, and architecture-driven compatibility. The result is a system that has been tested, validated, and incrementally built throughout its lifecycle ready for deployment with confidence in its performance, compliance, and mission alignment.

This ideal integrated technical workflow represents more than a modernization of engineering practices; it is a strategic re-architecture of how the DoD delivers capability. By linking MOSA, MBSE, Agile, DevSecOps, Digital Twins, and Test & Evaluation into a cohesive pipeline, the DoD can transition from fragmented delivery to continuous integration, validation, and value realization. Each component reinforces the others, creating a resilient, closed-loop system that thrives on adaptability, shared data, and unified intent. In an era where the speed of need defines success, this workflow is not optional, it is essential.

3.1.2 AI Augmented Enabled Pipeline

The future of defense capability delivery is not only digital, modular, and agile it is intelligent. Generative AI unlocks a new dimension in the DoD's technical workflow by augmenting human expertise with automated generation, optimization, and reasoning capabilities. Instead of merely accelerating existing processes, Generative AI delivers a step-change in the scale, adaptability, and responsiveness of the entire engineering value stream.

AI-Driven Design: Intelligent Modeling and Architecture

In this new ideal workflow, Generative AI assists in modeling mission threads directly from operational concepts, automatically creating MBSE artifacts such as behavior diagrams, use case flows, and architectural constraints. These models feed into a MOSA-compliant architecture, ensuring that system designs are inherently modular, interoperable, and upgradable. This frees system engineers from the manual translation of intent into artifacts, allowing them to focus on higher-value work like validating trade-offs and analyzing emergent behavior.

AI-Accelerated Delivery: Intelligent Development and Assurance

Simultaneously, Generative AI supports both software and hardware development by providing the scaffolds pf the components, and even recommends integrations with pre-existing libraries drastically reducing sprint time and enabling developers to focus on higher-level logic and system integration. In parallel, the AI generates a comprehensive suite of test scenarios mapped directly to safety, cybersecurity, and compliance objectives, which are then run continuously within the DevSecOps pipeline.

AI-assisted static analysis and runtime security checks can verify compliance in near real-time.

AI-Powered Validation: Intelligent Digital Twins and Integration

This entire flow is executed and validated within a digital twin environment, where the AI-generated code and models are simulated against real-world constraints, mission data, and user behavior. These insights allow developers and operators to rapidly adjust configurations and performance characteristics before anything is deployed to physical platforms. Meanwhile, hardware and software development processes are synchronized and integrated early via automated pipelines and shared models, removing the latency caused by legacy phase gates or hardware-software mismatches.

Ultimately, this results in continuous integration and deployment of validated, mission-ready capabilities through a modular architecture that evolves as fast as threats do. By embedding Generative AI across the lifecycle modeling, development, testing, and validation the DoD moves beyond being merely digital to become adaptive by design transforming speed, scale, and precision into a sustainable strategic advantage.

3.1.3 Agent Enabled

As the Department of Defense (DoD) moves beyond integrating Generative AI, the next frontier is Agentic AI systems capable of autonomous reasoning and orchestrating complex tasks. These are not passive tools but active teammates in engineering, operations, and mission planning. To support these agents, the technical pipeline must evolve from being merely automation-ready to being *human–machine teaming-native*. This requires a profound architectural shift: we must design our digital infrastructure for machine team members, not just human ones.

In an Agentic AI-enabled workflow, the user of engineering models and digital twins is an intelligent agent that must understand and adapt these systems autonomously. For this to be possible, components must expose clear APIs, maintain consistent data ontologies, and conform to interoperability standards. These are not just good engineering practices, they are non-negotiable requirements. The integrity of these requirements cannot be left to chance; it must be proven through formal methods that mathematically verify correctness, for example, by generating policy-proven code to ensure predictable behavior when agents interact.

MOSA principles become even more critical, as machine-readable contracts and interfaces allow agents to evaluate compatibility and trigger reconfiguration autonomously, with a high degree of confidence enabled by formally verified specifications. MBSE models must be structured so that agents can consume and refine them without constant human translation. Similarly, DevSecOps pipelines must be orchestratable by agents, allowing for continuous optimization based on mission context or emergent risk.

In such a system, operating on a foundation of formally verified components, an Agentic AI may autonomously generate a mission thread, design a configuration, validate it in simulation, and deploy it via the pipeline. The human role evolves to lead these human–machine teams shifting from managing discrete tasks to defining commanders' intent, setting strategic boundaries, and overseeing high-level decision loops. This approach is a logical extension of current trends that demands we build systems today with the expectation that the next operator may be a machine.

Without standardized APIs, common data models, and formally verifiable architectures, this level of autonomous orchestration is impossible. Legacy black box systems and proprietary silos will not scale in a world where agility must be machine-executed. Engineering for this future not only unlocks new capabilities it is the very foundation of relevance in a machine-accelerated pipeline.

3.2 Aligning People, Culture, and Policy

To deliver on future approach capabilities in an age defined by autonomy and machine-speed operations, the Department of Defense (DoD) must look beyond purely technical modernization. The success of modern practices from MOSA and MBSE to Agile and DevSecOps AI is ultimately gated not by the technology itself, but by the sociotechnical ecosystem: the people, policies, and culture that govern how value is delivered. Achieving this alignment requires a deliberate transformation across three critical domains**:** empowering mission-driven teams, mastering human-agent teaming, and fundamentally reforming policy.

3.2.1 Mission Empowered Teams Building the Foundation

Creating mission-oriented, empowered teams is the cornerstone of a resilient and adaptive delivery ecosystem. Such teams must be trusted to make real-time decisions aligned with mission outcomes, free from the constraints of hierarchical bottlenecks and compliance-driven inertia.

Key Elements of Empowered Team Formation:

- **Training:** Teams must cultivate deep expertise that is both broad and specific.
  - develop cross-functional fluency in Agile, digital engineering, cybersecurity, and AI,
  - master modern practices including Iterative planning, Modular Design, and DevSecOps, and
  - maintain a deep, current understanding of the mission context and the end-user's operational environment.
- **Authorities:** Teams must be granted the autonomy to act.
  - provide clearly defined authority over technical and programmatic decisions (scope, schedule, architecture) within set mission boundaries,

  - delegate the power to adjust priorities, modify requirements, and deploy capabilities in direct response to user feedback or evolving threats.

- **Leadership Support:** Leaders must actively champion and protect these teams.
  - act as enablers by removing organizational barriers and shielding teams from legacy governance,
  - reshape performance incentives to reward the delivery of mission value and team outcomes, not the completion of compliance artifacts.

When teams are empowered and supported by leaders, they move from simply completing tasks to delivering mission critical business outcomes which is required to organizations to remain competitive in the digital age.

3.2.2 Human/Agent Teaming

As Generative and Agentic AI systems mature, effective human-agent teaming will become a decisive source of military advantage. This requires a fundamental shift: our teams must be trained not merely to use AI tools, but to co-create, supervise, and evolve alongside them.

Required Human-Agent Teaming Competencies:

- **Training Programs:** AI readiness must be woven into professional military education.
  - establish robust training in core AI concepts (machine learning, confidence metrics, human-in-the-loop models) for warfighters, engineers, and acquisition professionals alike,
  - develop operator-focused curricula centered on interpreting AI outputs, managing escalation pathways, and applying counterfactual reasoning to AI recommendations.
- **New Leadership Models:** Command must be adapted for the age of autonomy.
  - prepare commanders to lead integrated teams of humans and autonomous agents,
  - develop critical trust calibration skills: knowing when to delegate to an AI and when a human must intervene, and
  - evolve leadership from the role of task allocator to commander's intent communicator, providing strategic guidance for autonomous systems to execute.
- **Operational Concepts:** Doctrine must keep pace with technology.
  - develop and codify new TTPs (tactics, techniques, and procedures) for integrating AI into mission threads,
  - establish clear AI failover protocols and battle rhythm synchronization with autonomous agents, and

- incorporate agent-based operational wargaming to test and refine human–machine interaction under stress.

Ultimately, human-agent teaming must become a core competency at every level of the force, blending technical acumen with operational judgment to harness the power of autonomy while ensuring unwavering human control and accountability.

3.2.3 Process Reform in Action

Even the most Agile, AI-integrated technical workforce will be throttled by outdated acquisition and oversight policies. To support continuous learning and adaptive systems, the DoD's governance itself must be reimagined to match the speed and flexibility of modern practices.

Actionable Policy Reforms Include the Following

- **Modern Contracting Vehicles**
  - expand the use of Other Transaction Authorities (OTAs), Consortium-Based Acquisition, and Software Acquisition Pathways to accelerate contracting,
  - structure contracts modularly to fund and authorize work in alignment with Agile increments and DevSecOps pipelines, allowing for evolving requirements.
- **Iterative Budget Alignment**
  - shift from rigid, annual milestone-based funding to more fluid models like rolling-wave planning and multi-year, increment-based appropriations,
  - align Planning, Programming, Budgeting, and Execution (PPBE) reform with delivery cadence (e.g., quarterly deployment planning) to synchronize funding with value creation.
- **Ethics and Oversight for Learning Systems**
  - establish standing oversight boards for real-time validation of AI behavior against commander's intent,
  - define clear policy guardrails for adaptive AI, including operational boundaries, data provenance standards, and auditability-by-design to ensure accountability.
- **Continuous Validation Frameworks**
  - implement digital governance infrastructure that automates the traceability of AI behavior back to validated requirements,
  - use DevSecOps telemetry, digital twin simulations, and operational feedback loops to create a living audit trail for continuous assurance and accreditation.

These reforms are not merely bureaucratic adjustments; they are the foundational enablers of the speed and adaptability the DoD requires. By reshaping policy to be as modular and iterative as the technologies it governs, the Department can finally close the sociotechnical divide and unlock the full potential of its empowered, AI-augmented teams. To fully realize the promise of advanced digital and autonomous systems, the DoD must transform not just its technical tooling, but its human, organizational, and policy foundations. Mission-empowered teams, human-agent teaming, and policy reform are not peripheral enhancements; they are core pillars of a future-ready sociotechnical system. The next generation of warfighting capabilities will not be built by tools alone but by teams and institutions aligned in purpose, equipped with modern authority, and prepared to collaborate with machines in service of national defense.

4 A Plan for Change

4.1 Phase 1 Foundations Establishing the Core

4.1.1 Technical Integration

The foundation phase is critical for establishing a unified technical delivery engine that is resilient, adaptable, and secure from the outset. This initial phase focuses on four key requirements that mitigate risk early and maximize development velocity.

- **Toolchain Integration** Begin by consolidating disjointed engineering, development, testing, and compliance tools into a single, cohesive pipeline. This integration breaks down departmental silos and creates a single source of truth, dramatically reducing friction and error rates while accelerating workflow.
- **Modular Architecture** A Modular Open Systems Approach (MOSA) is adopted to ensure all hardware and software components are interoperable and easily upgradeable. This approach prevents vendor lock-in and future-proofs the system, allowing for rapid, cost-effective technology insertion as new innovations become available.
- **Security by Design** Security is integrated from the start by embedding shift-left practices directly into the development and testing processes. Using tools like AI-enabled threat detection and static code analysis, vulnerabilities are identified and remediated at the earliest stages. This proactive stance drastically reduces the future cost and complexity of security compliance while ensuring a more resilient final product.
- **Digital Thread Enablement** Finally, model-centric practices are used to establish a complete digital thread, creating end-to-end traceability from concept through code. This provides a fully transparent and auditable record of the project lifecycle, which simplifies compliance checks and makes impact analysis for any proposed changes both fast and precise.

4.1.2 Sociotechnical Integration

With the technical engine in place, the next phase focuses on cultivating the human and procedural ecosystem required to operate with true agility. This involves a deliberate focus on creating a common culture, empowering teams, evolving leadership, and reforming procurement.

- **Cross-Functional Training** Shared learning sessions are launched across all key disciplines including Program Management, Systems Engineering, Software Development, Hardware Development, Cybersecurity, Test, and Operations to build a common lexicon. This foundational step breaks down communication barriers, eliminates jargon-based misunderstandings, and fosters the unified culture necessary for seamless collaboration.
- **Empowered Product Teams** Initial cross-functional teams are formed and empowered to deliver end-to-end value. This structure moves beyond traditional, siloed task completion, giving each team full ownership of a specific mission outcome. This fosters accountability, accelerates decision-making, and ensures development work remains tightly aligned with strategic goals.
- **Leadership Alignment** Leadership is aligned with this new way of working through targeted training in systems thinking, servant leadership, and Agile governance. This critically shifts the role of a leader from a directive manager to a strategic enabler who removes impediments and cultivates an environment where empowered teams can thrive.
- **Contract Reform Pilots** The operating model is supported by piloting flexible contract mechanisms, such as Other Transaction Authorities (OTAs). Because traditional, rigid contracts are a primary obstacle to agile development, these flexible vehicles provide the contractual agility needed to adapt to changing requirements and incorporate user feedback, ensuring the final product delivers maximum value.

Together, these initiatives create a cultural framework where empowered teams can leverage the new technical foundation to deliver continuous value.

4.2 Phase 2 Scaling, Expanding Across the Programs

4.2.1 Technical Integration

Horizontal Expansion of Integrated Pipelines

Original Expand integrated toolchains and DevSecOps practices across multiple programs, enabling reuse of libraries, components, and compliance frameworks.

Suggested Enhancement Frame this as a strategic move to create an internal platform.

Example This initiative expands integrated toolchains across multiple programs. The goal is to create an internal "platform" of reusable libraries, components, and automated compliance frameworks, allowing new programs to launch with unprecedented speed by eliminating the need to reinvent foundational capabilities.

AI-Augmented Engineering

Original Deploy Generative AI and Agentic AI for code generation, model validation, test case generation, and intelligent backlog grooming.

Suggested Enhancement Explain how this elevates the role of the human engineer.

Example Generative and Agentic AI are deployed to automate high-effort tasks like code generation, model validation, and test case creation. This deliberately augments human engineers, freeing them from routine work to focus on complex problem-solving and innovation, thereby dramatically increasing the organization's overall engineering velocity and creative capacity.

Digital Twins at Scale

Original Deploy digital twins for major subsystems to support predictive maintenance, simulation, and validation across system life cycles.

Suggested Enhancement Emphasize the benefit of continuous, virtual validation.

Example Digital twins are deployed for all major subsystems. This moves beyond simple modeling to create high-fidelity virtual environments for continuous validation, enabling predictive maintenance and robust simulation. It fundamentally shortens the system feedback loop from months or years to mere days, drastically reducing the reliance on costly physical testing.

Federated Data Standards

Original Implement shared data models and APIs to ensure interoperability across platforms and programs.

Suggested Enhancement Describe the benefit of treating data as a strategic, interoperable asset.

Example Shared data models and standardized APIs are implemented across the enterprise. This ensures data is treated as a strategic, interoperable asset not a trapped byproduct of individual programs unlocking advanced, cross-program analytics and enabling a holistic, real-time view of the entire system-of-systems.

4.2.2 Sociotechnical Integration

Scaling Mission-Oriented Teams

Original Institutionalize empowered teams across programs with embedded compliance, cyber, and test personnel.

Suggested Enhancement Highlight how this makes security and compliance an organic part of the process.

Example The empowered team model is institutionalized across all programs, with compliance, cyber, and test expertise embedded directly within each team. This makes accountability for quality and security an organic function of the development process, eliminating the delays and friction of external handoffs.

Governance Evolution

Original Create cross-domain governance forums to ensure technical and organizational alignment.

Suggested Enhancement Clarify that this is about just enough governance, not reintroducing bureaucracy.

Example Lightweight, cross-domain governance forums are created. These forums provide "just enough" oversight to ensure strategic and technical alignment across teams without reintroducing the bureaucratic bottlenecks of traditional governance, enabling scalable operations without sacrificing agility.

Frozen Middle Mobilization

Original Redesign mid-level management roles to support coaching, strategy alignment, and value delivery instead of command-and-control oversight.

Suggested Enhancement Directly address the common frozen middle challenge and its solution.

Example This initiative directly addresses the common "frozen middle" challenge by redesigning mid-level management roles. The focus shifts from command-and-control oversight to coaching, strategy alignment, and removing obstacles, transforming this critical layer from a potential source of resistance into a powerful catalyst for change.

Workforce Upskilling

Original Establish enterprise-wide AI fluency programs and human-agent teaming certifications.

Suggested Enhancement Position this as future-proofing the organization's workforce.

Example Enterprise-wide AI fluency programs and human-agent teaming certifications are established. This future-proofs the organization's most asset its people ensuring the workforce is not just equipped with new tools but is culturally and professionally prepared for a future where human-AI collaboration is central to value creation.

4.3 Phase 3 Adaptive Enterprise

4.3.1 Technical Integration

Enterprise-Wide Digital Backbone

Original Standardize digital thread and DevSecOps infrastructure across the organization with compliance, traceability, and security as defaults.

Suggested Enhancement Frame this as the ultimate enterprise utility, like a power grid for innovation.

Example The digital thread and DevSecOps infrastructure are fully standardized into an enterprise-wide digital backbone. This utility treats security, compliance, and traceability as non-negotiable, built-in services. It ensures that new projects can securely "plug in" and focus immediately on mission innovation rather than rebuilding foundational infrastructure.

Agentic AI Integration

Original Assume the user is increasingly an Agentic AI. Systems must now be built with robust APIs, real-time telemetry, and embedded policy enforcement. Testing, deployment, and operations pipelines must integrate AI feedback loops.

Suggested Enhancement Emphasize this as a fundamental paradigm shift in system design.

Example A core paradigm shift is embraced: the user is now an AI Agent as often as a human. Consequently, systems are designed API-first and not' built not as passive tools but as active participants in a dynamic, AI-driven ecosystem. They feature robust APIs, real-time telemetry, and embedded policy enforcement to support machine-speed interaction and continuous AI feedback loops.

Autonomous System Certification

Original Embed continuous validation mechanisms and assurance cases within pipelines to support AI and autonomous capabilities post-deployment.

Suggested Enhancement Position this as the solution to trust at scale for AI.

Example The final evolution is a framework for certifying autonomous systems in real-time. By embedding continuous validation and dynamic digital assurance cases directly into the operational pipeline, the enterprise can perpetually verify that its AI capabilities are operating safely and effectively long after deployment. This solves the critical challenge of trust at scale, enabling the confident and responsible use of mission-critical autonomous systems.

4.4 Sociotechnical Integration

Policy and Budget Reform

Original Institutionalize iterative acquisition, milestone-independent funding, and ethics-driven oversight models capable of adapting to dynamic learning systems.

Suggested Enhancement Position this as breaking free from legacy bureaucracy that hinders agility.

Example Core bureaucratic structures are reformed by institutionalizing iterative acquisition and milestone-independent funding. This decouples progress from rigid, outdated plans, providing the financial flexibility to learn and adapt. It is complemented by new, ethics-driven oversight models designed specifically for the unique challenges of dynamic AI systems.

Operational Concepts for AI

Original Introduce doctrine and operational workflows that position human-agent teaming as a strategic advantage, including role definitions, command escalation paths, and situational risk assessment.

Suggested Enhancement Clarify that this moves AI from being a tool to a true teammate.

Example New doctrine and operational workflows are introduced to formally treat human-agent teaming as a core strategic advantage. This moves beyond simply using AI as a tool, establishing it as a true teammate with clearly defined roles, command escalation paths, and risk assessment frameworks, providing the clarity required for humans and AI to operate together effectively under pressure.

Cultural Shift Measurement

Original Track and celebrate cross-functional collaboration, decentralized decision-making, and AI trust benchmarks as new indicators of enterprise health.

Suggested Enhancement Emphasize that redefining success metrics is what makes the culture change real.

Example The metrics for success are redefined. The enterprise begins to track and celebrate collaboration rates, the speed of decentralized decisions, and AI trust benchmarks as leading indicators of organizational health. This makes the desired cultural shift tangible and powerfully reinforces that agility, trust, and empowerment are now core, measurable business values.

Institutionalize Learning

Original Embed retrospectives, knowledge sharing, and experimental pilots into everyday operations. Establish innovation as a cultural norm, not an exception.

Suggested Enhancement Frame this as the creation of a permanent learning organization.

Example Finally, a learning mindset is institutionalized by embedding retrospectives, knowledge sharing, and a 'safe-to-fail' approach to experimental pilots into the rhythm of everyday operations. This ensures that innovation is no longer an isolated event or the job of a special team but becomes the cultural norm and the default behavior across the entire enterprise.

5 Conclusion

In the face of mounting global threats and rapidly advancing technologies, the DoD finds itself at a strategic crossroads. It can no longer afford to pursue fragmented modernization efforts that apply new technologies within the confines of outdated bureaucratic and organizational structures. This chapter has argued that the DoD's greatest challenge is not technological, but sociotechnical: the integration of people, process, policy, and technology will ultimately determine the nation's ability to deliver mission-ready capabilities at the speed of need.

The rise of Artificial Intelligence particularly Generative and Agentic AI has accelerated the pace of technological change to such a degree that legacy systems of acquisition, governance, and cultural behavior are no longer merely inefficient; they are actively detrimental to mission readiness. AI is not just another tool in the DoD's arsenal it is a capability multiplier that demands a tightly integrated sociotechnical ecosystem. Without reforming the human, organizational, and policy context in which it operates, AI cannot be trusted, scaled, or fully realized. In this new era, speed itself has become the pacing threat and the ability to adapt with speed, not just deliver with precision, is the new form of overmatch.

Throughout this chapter, we have demonstrated that isolated adoption of methodologies like Agile, DevSecOps, MBSE, or MOSA is insufficient. When practiced in silos, these methods clash rather than converge, resulting in brittleness, misalignment, and missed opportunities. Programs fail not because the tools are ineffective, but because the broader system governed by hierarchical control structures, outdated contracting models, and legacy performance metrics is unable to absorb their benefits. The real transformation must therefore be systemic, deliberate, and deeply integrative.

What is required is a new operating model for the DoD: one that builds upon a unified digital backbone, powered by model-based engineering, modular architectures, continuous delivery pipelines, and AI-augmented workflows.

Yet even this technical foundation is only effective when it is supported by empowered teams, federated governance, adaptive leadership, and policies restructured around value delivery rather than documentation. The cultural, organizational, and regulatory reforms proposed across Phases 1 through 3 of these plans outline the pathway to becoming a truly adaptive enterprise: a defense institution that learns, iterates, and evolves faster than its adversaries.

To achieve this, the DoD must shift its identity from a system that delivers projects to one that learns continuously. It must move from a culture of compliance to a culture of curiosity; from rigid oversight to resilient autonomy; from waterfall gatekeeping to digital feedback loops. Success will not be defined by how many systems are delivered, but by how quickly teams can learn, adapt, and deploy mission-ready capabilities in response to emerging threats.

In this light, the integration of technology is not the destination, it is the starting point. The true destination is an enterprise where every layer of the sociotechnical system is aligned for speed, trust, and resilience. Where empowered human-agent teams operate in harmony with AI to achieve mission outcomes. Where contracts and policies enable agility rather than impede it. And where cultural transformation is not an afterthought, but the foundation of sustainable change.

The next frontier of defense superiority will not be won solely through platforms, systems, or budgets. It will be won by organizations that can adapt faster than the environment around them. In this new era, the only true overmatch is the speed of adaptation. A unified sociotechnical system is the engine of that speed. And the time to build it is now.

References

1. U.S. Government Accountability Office. (2024). Weapon systems annual assessment: DOD is not yet well-positioned to field systems with speed (Report No. GAO-24-106831). Washington, DC: Author. Retrieved October 5, 2025, from https://www.gao.gov/products/gao-24-106831
2. U.S. Government Accountability Office. (2025). Weapon systems annual assessment: DOD leaders should ensure that newer programs are structured for speed and innovation (Report No. GAO-25-107569). Washington, DC: Author. Retrieved October 5, 2025, from https://www.gao.gov/products/gao-25-107569
3. U.S. Government Accountability Office. (2023). Weapon systems annual assessment: Programs are not consistently implementing practices that can help accelerate acquisitions (Report No. GAO-23-106059). Washington, DC: Author. Retrieved October 5, 2025, from https://www.gao.gov/products/gao-23-106059
4. U.S. Government Accountability Office. (2025). F-35 joint strike fighter: Actions needed to address late deliveries and improve future development (Report No. GAO-25-107632). Washington, DC: Author. Retrieved October 5, 2025, from https://www.gao.gov/products/gao-25-107632
5. U.S. Government Accountability Office. (2024). F-35 joint strike fighter: Program continues to encounter production issues and modernization delays (Report No. GAO-24-106909). Washington, DC: Author. Retrieved October 5, 2025, from https://www.gao.gov/products/gao-24-106909
6. U.S. Government Accountability Office. (2025). F-35 joint strike fighter (Report No. GAO-25-107632). Washington, DC: Author. Retrieved October 5, 2025, from https://www.gao.gov/assets/gao-25-107632.pdf
7. U.S. Department of War. "Defense Innovation Board Holds Summer 2024 Meeting." Last modified July 17, 2024. https://www.war.gov/News/Releases/Release/Article/3841443/defense-innovation-board-holds-summer-2024-meeting/
8. Defense Innovation Board. (2024, July 1). Aligning incentives to drive faster tech adoption. Office of the Under Secretary of Defense for Research & Engineering. Retrieved October 5, 2025, from https://innovation.defense.gov/Portals/63/20240701%20DIB%20Report_Aligning%20Incentives%20PUBLISHED%20STUDY_1.pdf
9. Defense Innovation Board. *Defense Innovation Board Quarterly Meeting*. Washington, DC: Defense Visual Information Distribution Service, July 17, 2024. https://www.dvidshub.net/video/930833/defense-innovation-board-quarterly-meeting

10. Air & Space Forces Magazine. (2025, March 3). Air Force software factory Kessel Run is pivoting. Not everyone is happy. Retrieved October 5, 2025, from https://www.airandspaceforces.com/kessel-run-air-force-software-factor-pivoting/
11. Defense Acquisition University. (n.d.). Kessel Run: A "day in the life" at a DoD software factory. Retrieved October 5, 2025, from https://media.dau.edu/media/Kessel+RunA+a+%22Day+in+the+Life%22+at+a+DoD+Software+Factory/1_y9aexv7y
12. Air & Space Forces Magazine. (2022, April 29). Making the Kessel Run. Retrieved October 5, 2025, from https://www.airandspaceforces.com/article/making-the-kessel-run/
13. U.S. Government Accountability Office. *Defense Acquisitions: Assessments of Selected Weapon Programs (GAO-16-329SP).* Washington, DC: U.S. Government Accountability Office, 2016. https://www.gao.gov/products/gao-16-329sp
14. U.S. Government Accountability Office. *Defense Acquisitions: Assessments of Selected Weapon Programs (GAO-17-333SP).* Washington, DC: U.S. Government Accountability Office, 2017. https://www.gao.gov/products/gao-17-333sp
15. U.S. Government Accountability Office. *Weapon Systems Annual Assessment: Knowledge Gaps Pose Risks to Sustaining Recent Positive Trends (GAO-18-360SP).* Washington, DC: U.S. Government Accountability Office, 2018. https://www.gao.gov/products/gao-18-360sp
16. U.S. Government Accountability Office. *Weapon Systems Annual Assessment: Limited Use of Knowledge-Based Practices Continues to Undercut DOD's Investments (GAO-19-336SP).* Washington, DC: U.S. Government Accountability Office, 2019. https://www.gao.gov/products/gao-19-336sp
17. U.S. Government Accountability Office. *Defense Acquisitions Annual Assessment: Drive to Deliver Capabilities Faster Increases Importance of Program Knowledge and Consistent Data for Oversight (GAO-20-439).* Washington, DC: U.S. Government Accountability Office, 2020. https://www.gao.gov/products/gao-20-439
18. U.S. Government Accountability Office. *Weapon Systems Annual Assessment: Updated Program Oversight Approach Needed (GAO-21-222).* Washington, DC: U.S. Government Accountability Office, 2021. https://www.gao.gov/products/gao-21-222
19. U.S. Government Accountability Office. *Weapon Systems Annual Assessment: Challenges to Fielding Capabilities Faster Persist (GAO-22-105230).* Washington, DC: U.S. Government Accountability Office, 2022. https://www.gao.gov/products/gao-22-105230
20. U.S. Government Accountability Office. *Weapon Systems Annual Assessment: Programs Are Not Consistently Implementing Practices That Can Help Accelerate Acquisitions (GAO-23-106059).* Washington, DC: U.S. Government Accountability Office, 2023. https://www.gao.gov/products/gao-23-106059
21. Rendon, Rene G., et al. "Contracting processes and structures for systems-of-systems acquisition". Systems Engineering, vol. 15, no. 4, 2012, p. 471–482. https://doi.org/10.1002/sys.21214
22. International Council on Systems Engineering. Systems Engineering Book of Knowledge. Accessed Feb 28, 2022. Published online 2021. https://www.sebokwiki.org/wiki/Guide_to_the_Systems_Engineering_Body_of_Knowledge_(SEBoK)
23. Fazio, Sam, and Karen LaFond. Applying Agile to Mitigate the Risk of Transitioning Defense Embedded Software Technologies. Acquisition Research Program, 2025
24. Barkin, Ian, and Thomas H. Davenport. "Harnessing grassroots automation." MIT Sloan Management Review 65.1 (2023): 74–78.
25. Govers, Mark, et al. "A theoretical essay on socio-technical systems design thinking in the era of digital transformation". Gruppe Interaktion Organisation Zeitschrift Für Angewandte Organisationspsychologie (Gio), vol. 54, no. 1, 2023, p. 27–40. https://doi.org/10.1007/s11612-023-00675-8
26. Neill, Marlene, et al. "Functional silos, integration & encroachment in internal communication". Public Relations Review, vol. 43, no. 4, 2017, p. 850–862. https://doi.org/10.1016/j.pubrev.2017.06.009
27. Montori, Federico, et al. "An iot toolchain architecture for planning, running and managing a complete condition monitoring scenario." IEEE Access 11 (2023): 6837–6856.

28. Lopez, C. Todd, and Major Wes Shinego. "Modern Software Acquisition to Speed Delivery, Boost Warfighter Lethality." DOD News, U.S. Department of Defense, 10 Mar. 2025, Defense.gov, Article 4114775.
29. Lyle, Lt Col Dave "Sugar", USAF. Overcoming "the Silo Effect" in the Department of Defense. Strategic Multilayer Assessment Invited Perspective Series, NSI, Inc., 5 Aug. 2019. nsiteam.com, https://nsiteam.com/social/wp-content/uploads/2019/08/Lyle-IP-2Aug19-Final2R.pdf.
30. Wada, Kazuo. The Evolution of the Toyota Production System. Berlin, Heidelberg, Germany: Springer, 2020.
31. Richmond, Ann M. Improving Depot Repair Lead Time Using Lean Six Sigma. Diss. The George Washington University, 2023.
32. Fowler, Martin, and Jim Highsmith. "The agile manifesto." Software development 9.8 (2001): 28-35.
33. Aroune, Jenny, Robert Hollister, and Nathan Taylor. Kessel Run: An Analysis of the Air Force's Internal Software Development Organization. Diss. Monterey, CA; Naval Postgraduate School, 2019.
34. Kim, Gene, et al. The DevOps handbook: How to create world-class agility, reliability, & security in technology organizations. It Revolution, 2021
35. Sirota, Sara. "Platform One modernizing DOD software development to enable digital ops." Inside the Air Force 31.28 (2020): 1–5.
36. Haskins, Cecilia. "4.6. 1 A historical perspective of MBSE with a view to the future." INCOSE international symposium. Vol. 21. No. 1. 2011.
37. Holladay, Jon B., et al. "MBSE Infusion and Modernization Initiative (MIAMI):"Hot" benefits for real NASA applications." 2019 IEEE Aerospace Conference. IEEE, 2019.
38. Zimmerman, Philomena, et al. "Considerations and examples of a modular open systems approach in defense systems." The Journal of Defense Modeling and Simulation 16.4 (2019): 373–388.
39. Reed, Tim. "The Future Airborne Capability Environment (FACE): A Strategic Imperative for U.S. Military Competitiveness." Military Embedded Systems, 16 May 2024, Military Embedded Systems, www.militaryembedded.com/avionics/software/the-future-airborne-capability-environment-face-a-strategic-imperative-for-us-military-competitiveness.
40. Smith, Larry. "Shift-Left Testing." Dr. Dobb's Journal: Software Tools for the Professional Programmer 26.9 (2001): 56–58.
41. Kraft, Edward M. "The air force digital thread/digital twin-life cycle integration and use of computational and experimental knowledge." 54th AIAA aerospace sciences meeting. 2016.
42. Pham, Duc Truong, and Rosemary S. Gault. "A comparison of rapid prototyping technologies." International Journal of machine tools and manufacture 38.10–11 (1998): 1257–1287.
43. Defense Advanced Research Projects Agency, U.S. Department of Defense, n.d., www.darpa.mil/research/programs/gremlins
44. Robles Martinez, Pamela, Abdul W. Basit, and Simon Gaisford. "The history, developments and opportunities of stereolithography." 3D Printing of Pharmaceuticals (2018): 55–79.
45. Stewart, Jack. "The Navy Is Using 3D Printers to Turn Warships into Weapons." Popular Mechanics, 2 Jun. 2022, PopularMechanics.com/articles/military/a40577952/navy-3d-printing-spares-drone-parts/
46. Grieves, Michael, and John Vickers. "Digital twin: Mitigating unpredictable, undesirable emergent behavior in complex systems." Transdisciplinary perspectives on complex systems: New findings and approaches (2017): 85–113.
47. Kraft, Edward M. "The air force digital thread/digital twin-life cycle integration and use of computational and experimental knowledge." 54th AIAA aerospace sciences meeting. 2016
48. Buchanan, Bruce G. "A (very) brief history of artificial intelligence." Ai Magazine 26.4 (2005): 53-53.
49. Choudhury, Rizwan. "Project Maven: The Epicenter of U.S.' AI Military Efforts." Interesting Engineering, 2 Mar. 2024, www.interestingengineering.com/military/project-maven-the-epi center-of-us-ai-military-efforts.

50. Switzky, Lawrence. "ELIZA effects: Pygmalion and the early development of artificial intelligence." Shaw 40.1 (2020): 50-68.
51. Howell, Christopher L., Kimberly E. Manser, and Raghuveer M. Rao, editors. Synthetic Data for Artificial Intelligence and Machine Learning: Tools, Techniques, and Applications. Proceedings of SPIE Volume 12529, SPIE–International Society for Optics and Photonics, 1–3 May 2023, Orlando, FL. SPIE, 2023. SPIE Digital Library
52. Vasan, Anjana. "The Fascinating History of Agentic Artificial Intelligence." Medium, 18 Mar. 2025, medium.com/wonder-think/the-fascinating-history-of-agentic-artificial-intelligence-14c4f3c6b58e.
53. Farnell, Rich and Kira Coffey. "AI's New Frontier in War Planning: How AI Agents Can Revolutionize Military Decision-Making." Field Artillery 2025 E-edition, Line of Departure, 27 Mar. 2025.
54. Mishra, Alok, et al. "Structured software development versus agile software development: a comparative analysis". International Journal of System Assurance Engineering and Management, vol. 14, no. 4, 2023, p. 1504–1522. https://doi.org/10.1007/s13198-023-01958-5
55. Yeman, Robin. The Application of Agile to Large-Scale, Safety-Critical, Cyber-Physical Systems. Diss. Colorado State University, 2025
56. Porchia, Jamie, Daniel Finkenstadt, and Grant Wilson. Unpacking the Authority to Operate (ATO) Process: Implications for the DoD. Acquisition Research Program, 2025
57. Arifin, Firdaus, and Ihsanul Maarif. "Restructuring the Division of Governmental Functions Based on Lean Governance Principles." Journal Equity of Law and Governance 6.1 (2024): 164-174.
58. Model-Based Systems Engineering (MBSE)." Guide to the Systems Engineering Body of Knowledge (SEBoK), Systems Engineering Body of Knowledge, 2014, SEBoK Wiki. Accessed 8 June 2025.
59. Carroll, E. and Malins, R. (2016). Systematic literature review: how is model-based systems engineering justified?. https://doi.org/10.2172/1561164
60. U.S. Government Accountability Office. Weapon Systems Acquisition: DOD Needs Better Planning to Attain Benefits of Modular Open Systems. GAO-25-106931, 22 Jan. 2025
61. McKinsey & Company, "What is Digital-Twin Technology?" McKinsey.com (website), July 12, 2023. https://www.mckinsey.com/featured-insights/mckinsey-explainers/what-is-digital-twin-technology.
62. S. Johnson and R. Yeman, The Application of Industrial DevOps Using Digital Twins, IT Revolution, May 7, 2024. [Online]. Available: https://itrevolution.com/product/application-of-industrial-devops-using-digital-twins/. Accessed: Jun. 8, 2025
63. White, Morgan, and Steven Ellington. Adoption of Digital Twin Within the Department of the Navy. Diss. Acquisition Research Program, 2022.
64. Wang, Yuanbin, Robert Blache, and Xun Xu. "Selection of additive manufacturing processes." Rapid prototyping journal 23.2 (2017): 434–447.
65. Roper, Will. "There is no spoon: The new digital acquisition reality." Defense AR Journal 28.4 (2021): 488-488.
66. Arndt, Craig, Awele Anyanhun, and Jeremy S. Werner. Shifting left: Opportunities to reduce defense acquisition cycle time by fully integrating test and evaluation in model based systems engineering. Acquisition Research Program, 2023.
67. Miller, Andrew W., Ronald E. Giachetti, and Douglas L. Van Bossuyt. "Challenges of adopting devops for the combat systems development environment." Defense Acquisition Research Journal: A Publication of the Defense Acquisition University 29.1 (2022).
68. "How to Use Generative AI to Boost Developer Productivity." Gartner, https://www.gartner.com/en/software-engineering/insights/how-to-use-gen-ai-to-boost-developer-productivity. Accessed 8 June 2025.
69. Weisz, Justin D., et al. "Examining the use and impact of an AI code assistant on developer productivity and experience in the enterprise." Proceedings of the Extended Abstracts of the CHI Conference on Human Factors in Computing Systems. 2025.

70. Sawant, Prashant. “Agentic AI: A Quantitative Analysis of Performance and Applications.” (2025).
71. Farnell, Lt Col Rich, and Lt Col Kira Coffey. “AI’s New Frontier in War Planning.” Field Artillery—2025 E-Edition, Line of Departure, U.S. Army, 11 Oct. 2024, https://www.lineofdeparture.army.mil/Journals/Field-Artillery/Field-Artillery-Archive/Field-Artillery-2025-E-Edition/AIs-New-Frontier-in-War-Planning/. Accessed 8 June 2025.
72. Bergengruen, Vera. “How Tech Giants Turned Ukraine Into and AI War Lab.” TIME Magazine, February 8, 2024. https://time.com/6691662/ai-ukraine-war-palantir/
73. Gurov, Viktor, et al. “Quality management integration in the functionality of product lifecycle management in energy-effective production”. E3s Web of Conferences, vol. 460, 2023, p. 10034. https://doi.org/10.1051/e3sconf/202346010034
74. Sira, Mariya. “Artificial intelligence in knowledge capture and transfer: breaking down organizational silos.” (2024).
75. Sheridan, Arthur E., and Robert Burnes. “F-35 program history–from JAST to IOC.” 2018 Aviation Technology, Integration, and Operations Conference. 2018
76. U.S. Government Accountability Office. F-35 Joint Strike Fighter: Program Continues to Encounter Production Issues and Modernization Delays. GAO-24-106909, 16 May 2024
77. Meagan Metzger, “There isn’t just one Valley of Death: Tackling the DoD transition problem,” DefenseScoop, March 1, 2023.
78. Mitchell, Billy. “Lack of Understanding Holding Back Military AI Adoption, Leaders Say.” DefenseScoop, 29 Apr. 2025, https://defensescoop.com/2025/04/29/lack-of-understanding-holding-back-military-ai-adoption/. Accessed 8 June 2025.
79. Talbert Jr, James. How Acquisition Decisions Impacted Fleet Readiness in the LCS Platform. Diss. Acquisition Research Program, 2024.
80. U.S. Government Accountability Office. Weapon Systems Annual Assessment: DOD Is Not Yet Well Positioned to Deliver Warfighter Capabilities on Time and at an Affordable Cost. GAO-24-106831, July 18, 2024.
81. Wiggins, “China’s Rapid Rise in AI and Technology: What the World Can Learn,” LinkedIn, [Online]. Available: https://www.linkedin.com/pulse/chinas-rapid-rise-ai-technology-what-world-can-learn-dion-wiggins-btqoc/. [Accessed: 08-Jun-2025].
82. L. Darina, “How fast is technology growing statistics [Updated 2023],” Leftronic, Mar. 7, 2023. [Online]. Available: https://leftronic.com/blog/how-fast-is-technology-growing-statistics/. [Accessed: Jun. 4, 2025]
83. K. H. Hicks, “Remarks by Deputy Secretary of Defense Kathleen H. Hicks at the Defense Innovation Summit,” U.S. Department of Defense, transcript, Feb. 27, 2025. [Online]. Available: https://www.defense.gov/News/Speeches/Speech/Article/4028027/remarks-by-deputy-secretary-of-defense-kathleen-h-hicks-at-the-defense-innovati/. [Accessed: Jun. 8, 2025]
84. Defense Innovation Board, Aligning Incentives to Drive Faster Tech Adoption, Office of the Under Secretary of Defense for Research & Engineering, July 1, 2024. [Online]. Available: innovation.defense.gov/Portals/63/20240701 DIB Report_Aligning Incentives PUBLISHED STUDY_1.pdf. [Accessed: Jun. 3, 2025]
85. Ukraine’s defense tech is shaping the future of warfare,” Yahoo News, 2025. [Online]. Available: https://www.yahoo.com/news/ukraine-defense-tech-shaping-future-120724596.html. [Accessed: Jun. 8, 2025]
86. Heyman, K. (2024, June 6). Toward a software-defined hardware world. Semiconductor Engineering. https://semiengineering.com/toward-a-software-defined-hardware-world/
87. Lopez, C. T., & Shinego, W. M. (2025, March 10). Modern software acquisition to speed delivery, boost warfighter lethality. U.S. Department of Defense.
88. U.S. Government Accountability Office. (2024, December 12). Actions needed to address acquisition workforce challenges and data (GAO-25-107075). https://files.gao.gov/reports/GAO-25-107075/index.html
89. Dorton, Stephen, et al. “Adaptations to trust incidents with artificial intelligence”. Proceedings of the Human Factors and Ergonomics Society Annual Meeting, vol. 66, no. 1, 2022, p. 95–99. https://doi.org/10.1177/1071181322661146

90. U.S. Department of Defense. 2019. *DoD Instruction 5000.80: Operation of the Middle Tier of Acquisition.* Washington, DC: Department of Defense. PDF. https://www.esd.whs.mil/Portals/54/Documents/DD/issuances/dodi/500080p.PDF
91. Krimmelbein, Fred. "The Ethical Implications of AI and Job Displacement." Sogeti Labs, 3 Oct. 2024, labs.sogeti.com/the-ethical-implications-of-ai-and-job-displacement/.
92. ŞAHiN, Emrullah, Naciye Nur Arslan, and Durmuş Özdemir. "Unlocking the black box: an in-depth review on interpretability, explainability, and reliability in deep learning." Neural Computing and Applications 37.2 (2025): 859–965.
93. Maathuis, Clara. "Human-Centred AI in Military Cyber Operations." 19th International Conference on Cyber Warfare and Security: ICCWS 2024. Academic Conferences and publishing limited, 2024.
94. Vizitiu, Cristian, et al. "Optimizing early-stage efficiency in complex system development via model-based systems engineering (MBSE) and concurrent engineering (CE) integration." Research in Engineering Design 36.3 (2025): 1–36.
95. Huss, M., Herber, D. R., & Borky, J. M. (2023). An Agile Model-Based Software Engineering Approach Illustrated through the Development of a Health Technology System. Software, 2(2), 234-257. https://doi.org/10.3390/software2020011

Navigating Computational Challenges and Security Considerations in Defense Industry Modeling and Simulation

Aleksandra Markina-Khusid, Diane Villanueva, and Fallon Laliberte

Abstract Modeling and simulation (M&S) plays a critical role in the defense industry, enabling the analysis of complex systems and scenarios to inform decision-making and enhance system performance. This book chapter explores the role of conceptual models in M&S, addressing key challenges and considerations for developing and executing M&S frameworks in classified environments.

1 Introduction

Modeling and simulation (M&S) plays a key role in modern defense acquisition and operations, offering military organizations the ability to explore and address the challenges of contemporary warfare within secure and efficient virtual environments. Through the creation of digital representations of real-world systems and scenarios, M&S facilitates experimentation, evaluation, and strategic planning without the inherent risks or expenses of live exercises. Its diverse applications provide critical support for optimizing military preparedness and achieving tactical and strategic objectives.

M&S supports a wide range of defense sector activities [1]. For training and education, it provides realistic virtual environments where soldiers, pilots, and commanders can practice tactics, procedures, and decision-making. These simulations range from individual flight simulators to large-scale, networked wargames involving multiple units across different services. In acquisition and engineering, M&S facilitates virtual prototyping and testing of new weapon systems, vehicles, and equipment, significantly reducing development costs and risks [2]. For operational analysis and planning, M&S enables the evaluation of military capabilities, exploration of strategic options, and assessment of potential outcomes in complex scenarios [1].

A. Markina-Khusid (✉) · D. Villanueva
Systems Engineering, The MITRE Corporation, Bedford, MA, United States
e-mail: amarkina@gmail.com

F. Laliberte
Nuclear Enterprise, The MITRE Corporation, Bedford, MA, United States

A. Weissman (ed.), *Women in Advanced Computing and Software for Defense*, Women in Engineering and Science, https://doi.org/10.1007/978-3-032-17466-6_4

The primary motivation for employing M&S in defense lies in its ability to analyze highly complex and dynamic systems that are too expensive, dangerous, or impractical to test in the real world. Modern military operations require involved interactions among satellites, aircraft, ground troops, and naval fleets, all operating in rapidly changing environments. M&S provides a risk-free laboratory to explore "what-if" scenarios, optimize operational strategies, and develop robust contingency plans. For example, it can simulate how a new radar system would perform against an emerging enemy threat or how effective a specific combat strategy would be under a stressing scenario. This capability enhances preparedness, identifies vulnerabilities, and ensures forces are ready for a wide array of contingencies.

An effective M&S effort starts with the creation of a conceptual model, which serves as the foundational blueprint for simulations. Conceptual models simplify the complexity of real-world systems into manageable representations, focusing on salient features while omitting irrelevant details. This abstraction ensures computational feasibility and interpretability, transforming M&S from a technical exercise into a powerful tool for informed decision-making. Conceptual models also act as communication tools, fostering shared understanding among stakeholders, including developers and subject matter experts. By defining objectives, assumptions, and key components, they align simulation efforts with strategic goals, ensuring validity, credibility, and utility [3, 4].

The importance of M&S in defense is further underscored by its ability to address challenges such as multi-scale interactions, high-dimensional phenomena, and uncertainties. Conceptual models guide the selection of appropriate M&S techniques and resources, balancing computational complexity with analytical rigor. Techniques such as physics-based modeling, uncertainty quantification, and design of experiments are integrated to ensure simulations remain robust and reliable. As M&S continues to evolve, advancements in immersive technologies, cloud computing, and surrogate modeling promise to further enhance its impact, revolutionizing the way military organizations train, analyze, and operate in the twenty-first century.

2 Conceptual Models in M&S

Conceptual models serve as the foundational blueprint, bridging real-world systems (RWSs) with their virtual representations. They are critical for managing inherent complexities, systematically addressing uncertainties, and strategically guiding the selection of appropriate M&S techniques and resources in domains characterized by high complexity and dynamic environments, such as defense. A well-defined conceptual model ensures the validity, credibility, and utility of simulation studies, transforming M&S from an abstract technical exercise into a powerful tool for informed decision-making and strategic advantage.

Formally, the conceptual model is a collection of abstractions, assumptions, and descriptions that represent the RWS, its environment, and behaviors [5]. It is independent of specific software or implementation details, serving as the logical structure upon which the computational model is built [6]. By identifying the optimal level of abstraction, the conceptual model ensures computational feasibility and interpretability, avoiding overly detailed models that are data-starved or computationally intractable [7]. The conceptual model also acts as a communication tool, fostering shared understanding among stakeholders, including developers and subject matter experts [6]. This alignment prevents costly rework and ensures the simulation addresses the intended problem. While simplifying reality may seem counterintuitive, precision in simulation often comes from focusing on salient features rather than replicating every detail [7].

Conceptual models are necessary for addressing the complexities inherent in defense systems, which often involve interconnected relationships, high-dimensional phenomena, multi-scale challenges, and uncertainties. Modern military operations encompass complex interactions across temporal, spatial, and domain-specific scales, such as the coordination of ground troops, air forces, and naval fleets in dynamic environments [6]. Conceptual modeling simplifies these complexities by defining the appropriate level of abstraction, focusing on factors directly influencing objectives while deliberately omitting irrelevant details [8].

High-dimensional problems in defense arise from the interplay of numerous interdependent variables, such as the performance of radar systems, troop movements, and logistics under varying conditions. Conceptual models address this by prioritizing salient features, ensuring simulations remain computationally feasible and analytically interpretable [8]. Multi-scale and multi-domain challenges are particularly pronounced in defense operations, where forces operate across diverse environments and interact with complex systems like satellites, aircraft, and naval assets. Conceptual models integrate perspectives by fostering shared understanding among stakeholders, including military commanders and simulation developers. Frameworks like NATO's Conceptual Model of the Mission Space (CMMS) standardize representation across domains, enhancing interoperability and reusability of simulation assets [4].

Uncertainty is another critical challenge in defense M&S. Conceptual models explicitly document assumptions and simplifications, framing uncertainty as a design parameter rather than a limitation. This approach integrates uncertainty quantification into simulation design, ensuring transparency and adaptability [4, 6]. Iterative refinement further allows models to evolve as new data becomes available or objectives shift [8].

The conceptual model serves as a strategic foundation for determining the most suitable modeling techniques and resource allocation in simulation studies. By defining the system's objectives, scope, assumptions, and level of abstraction, it ensures alignment between the problem's intrinsic characteristics and the chosen simulation methodology. This technique-agnostic approach prevents tool-driven modeling, optimizes data requirements, and informs software and hardware resource decisions, ultimately enhancing the validity, feasibility, and utility of the simulation.

Table 1 Conceptual model's influence on M&S technique and resource selection

Conceptual model characteristic or decision	Impact on modeling technique	Impact on data requirements	Impact on software and hardware resources
Clear objectives or research question	Guide selection of techniques best suited to answer specific questions (e.g., process optimization vs. emergent behavior analysis) [5]	Focus data collection on relevant parameters for objectives [3]	Optimize resource use by avoiding unnecessary complexity [7]
Level of abstraction	Determine if a high-level (system dynamics) or detailed (ABM, DES) representation is needed [5]	Influence the volume and granularity of data required; simpler models need less data [5]	Directly affect computational complexity; "far abstraction" leads to simpler, less resource-intensive models [5]
Assumptions and simplifications	Can justify using a simpler technique or omitting complex interactions [7]	Reduce the need for extensive or hard-to-obtain data [7]	Decrease computational load and development time [7]
Multi-scale and multi-domain integration	May necessitate hybrid approaches or specialized frameworks for semantic consistency [4]	Reduce the need for extensive or hard-to-obtain data [7]	Decrease computational load and development time [7]

Table 1 describes how conceptual model guides selection of M&S techniques and resources.

3 Key Modeling and Simulation Techniques

In this section, key modeling and simulation techniques are described. Physics-based modeling, uncertainty quantification, and design of experiments are described. The goal is to provide the M&S practitioner with an overview of how to combine these techniques to ensure M&S assets are in line with the conceptual model laid out in the previous section.

3.1 Physics-Based Modeling

Physics-based modeling approaches use principles of physics to simulate and predict the behavior of systems. It involves creating mathematical models based on first principles—physical laws, such as Newton's laws of motion, thermodynamics, fluid dynamics, and quantum mechanics, to describe how systems evolve over time or respond to external influences. Because they rely on physical laws rather than data or approximations alone, physics-based models can offer more explainable predictions that help in determining accuracy.

However, real-world systems in the defense industry can be highly complex, requiring models that are similarly complex and computationally expensive. For dynamical systems, physical systems are represented with differential equations to describe relationships between force, energy, velocity, temperature, and displacement. To solve these computational models, a significant amount of processing power is required. Alternatively, simplifications to the model, including approximations or data-driven surrogate models, can make the problem tractable, but introduce errors in the model.

3.2 Uncertainty Quantification

Uncertainty quantification (UQ) plays a critical role in modeling and simulation within the defense industry, where the stakes are high, and decisions often have significant consequences. The ability to account for uncertainties in M&S ensures that predictions are robust, reliable, and capable of supporting informed decision-making under complex and dynamic conditions.

Uncertainty in modeling and simulation can be broadly categorized into two types: aleatory and epistemic. Aleatory uncertainty refers to inherent variability, such as material properties or environmental conditions, which are random and irreducible. Epistemic uncertainty arises from a lack of knowledge and can often be reduced through additional research, testing, or expert consultation [9, 10]. For example, epistemic uncertainty often occurs due to limitations in modeling complex phenomena, such as flow fields for aerodynamic modeling hypersonic weapons.

Forward propagation of uncertainty is a critical process in predictive modeling and simulation. In forward propagation, input parameter uncertainty is propagated through a computational model to affect the output quantities of interest (QoIs) as shown in Fig. 1.

Uncertainty in output QoIs can arise from multiple sources, including:

- **Input Parameter Uncertainty**: Variability in material properties, geometric parameters, or external loading conditions can propagate through the model, influencing the outputs.
- **Modeling Errors**: Errors in the computational model itself, such as simplifications or numerical approximations, can introduce additional uncertainty.

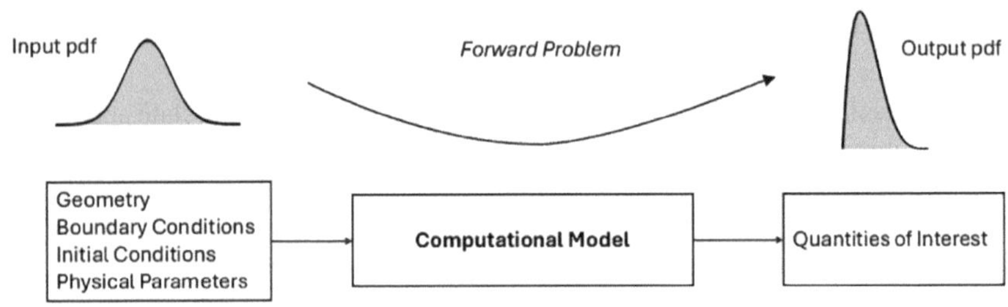

Fig. 1 Forward propagation of uncertainties through the computational model resulting in uncertainty in the output QoI

- **Simulation Noise**: Random fluctuations in simulation data can further contribute to uncertainty in the outputs.

As described in [11], capturing the uncertainty in output QoIs is essential for several reasons:

- **Multidisciplinary Systems**: In coupled systems, such as fluid-thermal-structural interactions in aerospace applications, uncertainty in the outputs of one discipline can serve as inputs to another, necessitating consistent and accurate representation.
- **Simulation Pipelines**: Maintaining consistency in uncertainty characterization prevents computational bottlenecks when outputs are used downstream in a simulation pipeline.
- **Traceability**: Preserving historic data, including uncertainty information, is crucial for traceable decision-making.
- **Future Reuse**: Accurate uncertainty characterization enables the reuse of data for new designs and applications, saving computational resources and time.

Calibration or inference is a statistical approach used to calibrate the uncertain parameters of computational models with observed or higher fidelity data as shown in Fig. 2.

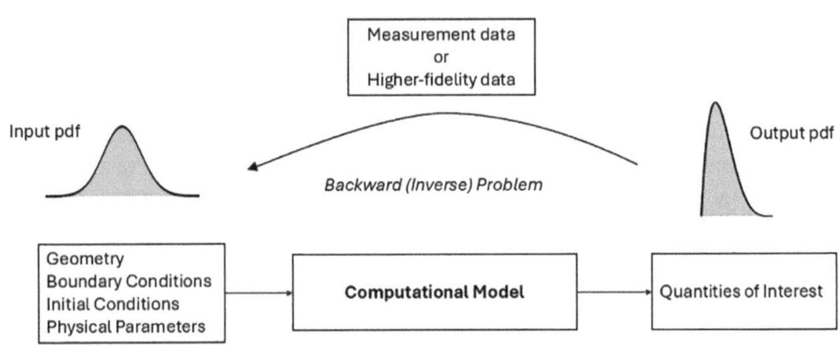

Fig. 2 Calibration of uncertain input parameters using observed data

Bayesian calibration is often used to update the probability distribution of uncertain parameters based on observed data. It is rooted in Bayes theorem, which is expressed as

$$P(y) = \frac{P(y|\theta)P(\theta)}{P(y)},$$

where $P(y)$ is the posterior distribution of the uncertain parameters θ given the observed data y, $P(\theta)$ is the likelihood of observing the data given the uncertain parameters, $P(\theta)$ is the prior distribution of the uncertain parameters, and $P(y)$ is the overall probability of observing y.

Several methods exist for Bayesian calibration and the choice is usually a trade-off between accuracy and computational cost [12–15]. Markov Chain Monte Carlo (MCMC) methods are sampling-based approaches that can describe complex posterior distributions [16]. However, since MCMC is a sampling-based method, many samples can be required to accurately obtain an accurate posterior distribution. Variational inference approaches are generally more efficient but require the assumption of the underlying family of distribution, such as the standard normal distribution [17]. Another well-studied method is to use transport maps to represent the posterior distribution as samples from a known distribution and use an optimization-based approach to learn the transport map parameters [18]. As discussed in [11], the choice of calibration method should be fit-for-purpose given the purpose of the model, amount of observational data, and the dimensionality.

3.3 Design Under Uncertainty

Designing systems and missions under uncertainty requires an understanding of both aleatory and epistemic uncertainties. Deterministic design methods use safety factors, safety margins, and knockdown factors are employed to account for uncertainties. These factors are typically based on historical data, experience, and regulatory requirements and are popular in aerospace applications [19]. The deterministic approach assumes that uncertainties can be compensated by conservative measures.

For example, in deterministic design, the maximum allowable temperature of a thermal protection system of a space vehicle might be set conservatively, ensuring that the design remains safe under worst-case scenarios [20]. This approach, while simple and intuitive, often leads to over-designed systems that are unnecessarily costly and heavy. As noted in [21], deterministic design may not lead to a minimum-cost design because it does not optimally allocate risk among failure modes.

In contrast, probabilistic design explicitly incorporates uncertainties into the design process by quantifying variability and errors. In reliability-based design optimization (RBDO), risk is optimally among failure modes, allowing designers to trade-off cost and performance while meeting specified reliability levels. The equation below presents a formulation of an RBDO problem that seeks to minimize the expectation of a system performance parameter $f(d, r)$ (e.g., the maximum temperature) such that the system level probability of failure $P_{f,sys}$ is less than the allowable P_f^{sys}.

$$E\big[f(d, r)\big] s.t. P_{f,sys}(d) \leq P_{f,sys}^{allow}.$$

A key aspect of probabilistic design is the integration of uncertainty reduction measures (URMs), such as tests and redesign, into the design process. As highlighted in [21], tests can reduce epistemic uncertainty by calibrating computational models and identifying dangerous or overly conservative designs. Redesign based on test results can restore safety margins or improve performance.

Probabilistic design offers several advantages over deterministic design:

- **Optimal Risk Allocation**: By quantifying uncertainties, probabilistic design allows risk to be allocated to the most difficult failure modes to protect against, while less risk is allocated to easier-to-protect modes.
- **Performance versus Cost Trade-Off**: Probabilistic design enables designers to balance performance and cost by considering the effect of uncertainties on reliability.
- **Reduction of Over-Design**: Unlike deterministic design, probabilistic design avoids unnecessary conservatism, leading to lighter and more cost-effective designs.

3.4 Design of Experiments

Design of experiments (DoE) is a method to plan and execute computational or physical experiments. This section focuses on methods that aid model calibration, uncertainty quantification, and prediction of targeted events such as system failures or operational limits. DoE enables the systematic exploration of design spaces to calibrate models, validate predictions, and identify critical regions of interest.

In defense modeling and simulation, accurate predictions of system behavior under extreme conditions are essential. Examples include predicting the onset of structural failure, aerodynamic instabilities, or thermal limits in hypersonic flight.

Several experimental design approaches are commonly used in defense applications:

1. **Alphabetic Design Criteria**: Alphabetic criteria, such as A-optimal and D-optimal designs, are widely used in experimental design for linear models. These criteria focus on minimizing specific measures of uncertainty in model predictions or parameter estimates [22]. For example, the D-optimal criterion maximizes the determinant of the Fisher information matrix, which corresponds to minimizing the volume of the confidence ellipsoid for the parameter estimates through maximizing the determinant of the Fisher information matrix.

2. **Factorial Designs (FF)**: These designs involve testing combinations of input variables at discrete levels. Full factorial designs provide comprehensive coverage of the design space by examining all combinations of levels for all variables; they can be computationally expensive, especially for high-dimensional problems [23].
3. **Space-Filling Designs**: Often used for computational experiments, space-filling design, such Latin Hypercube Sampling, seeks to provide spread data points in the design space to efficiently cover the space [24]. This approach is useful when the number of data points is limited due to cost.
4. **Sequential Experimental Designs**: Sequential designs iteratively add data points based on the information gained from previous experiments. This approach is also useful when the number of experiments is limited due to cost constraints.
5. **Targeted Designs for Events of Interest (EoI)**: For applications where specific events such as failure or transition are critical, targeted designs focus on regions near the limit states of interest. The Targeted Information Gain for Error Reduction (TIGER) method balances global model accuracy with local accuracy near the EoI [25].

The TIGER methodology is particularly suited for defense applications where both global accuracy and the prediction of targeted events are required. TIGER combines two objectives:

- **Global Accuracy**: Ensures the model is accurate across the entire design space by maximizing unweighted EIG.
- **Targeted Accuracy**: Focuses on regions near the EoI by weighing EIG with the probability of being near the limit state of the failure region.

Consider the case of predicting flutter boundaries and critical limit cycle oscillation (LCO) amplitudes for a panel in hypersonic flow as described in [25]. The TIGER method was used to identify the Mach number and flow conditions leading to flutter or critical LCO. The process involved:

- Building reduced-order models and surrogates for structural response and aerodynamic pressure.
- Using Gaussian Naïve Bayes classifiers to predict the probability of flutter and critical LCO.
- Iteratively selecting Mach numbers for the next experiment based on TIGER criteria.

Results demonstrated that TIGER effectively placed points near the limit states, improving the accuracy of predictions while maintaining global model fidelity.

While advanced DoE techniques offer significant advantages, their implementation requires careful consideration of computational costs, model complexity, and experimental budgets. Future research could explore hybrid approaches that integrate cost constraints into the design process.

4 Computational Cost Reduction Strategies

4.1 Methods to Reduce Computational Costs in M&S

4.1.1 High-Performance Computing

High-performance computing (HPC) can be leveraged to reduce computational costs for modeling and simulations (M&S), by enabling more efficient use of computational resources and staff time. HPCs can be used to distribute calculations across multiple processors or nodes, thereby shortening the time to run large-scale simulations. In particular, HPC utilization allows minimizing the need for multiple rounds of individual simulations to optimize various simulation parameters. Parallel processing not only accelerates time to results but also reduces energy, staff time, and hardware costs associated with prolonged computation on less powerful systems.

4.1.2 Problems Requiring Parallel Computing with HPC

HPCs are best suited for tasks that are computationally intensive and can utilize parallel processing. Parallel computing is beneficial for problems that involve large datasets, complex computations, or tasks that can be divided into independent or semi-independent subtasks. Parallel computing is often used to improve efficiency in modeling and simulation of large-scale problems, such as weather forecasting, or machine learning algorithms with large datasets. Parallel computing on HPCs also benefits situations in which engineers must perform overnight computational processing due to multiple design cycles or high-fidelity multi-physics simulations. These problems often require immense processing power and memory. An HPC can improve efficiency and reduce computation cost in these situations, by distributing the workload across multiple processors or nodes.

Radiation transport modeling with Monte Carlo N-Particle (MCNP) radiation transport code is one example of an area of work that benefits from parallelism on an HPC. Radiation transport modeling seeks to simulate the movement and interactions of radiation through materials. MCNP is an export-controlled general-purpose 3D radiation transport code that was developed and is maintained by Los Alamos National Laboratory [26]. MCNP uses random sampling of cross-section data, physics models, and radiation source descriptions to simulate the interactions of radiation with matter. MCNP has a variety of applications, ranging from radiation treatment planning in medical physics, advanced reactor design, and defense. Monte Carlo-based radiation transport codes inherently benefit from the use of an HPC, due to the parallel nature of the computational algorithm [27]. MCNP is designed to be parallel and scalable by allowing users to specify the number of threads and processors to run a simulation, via message passing interface (MPI) and OpenMP. In particle parallelism, the simulation distributes individual Monte Carlo runs, known as histories, to different compute nodes. Compute nodes then analyze the histories concurrently using the specified number of threads. MPI is commonly used in MCNP to distribute the problem across processors on an HPC and collect the tallied results, while OpenMP is used for threading computation on each individual processor [27]. The speedup of Monte Carlo radiation transport simulations using particle parallelism is limited by the time for communication and synchronization between processors [27]. Minimizing the rendezvous time by improvement in system hardware can result in a reduction in simulation time on the HPC. Another method Monte Carlo radiation transport systems use on HPCs to improve efficiency is called domain decomposition. Engineers running radiation transport simulations may need to do so with a Computer Aided Design (CAD) model. A mesh is required to represent the geometry and track the movement and interaction of radiation throughout the CAD model. Depending on the scale and complexity of the CAD (number of interfaces, small parts), this can result in a high demand in memory for MCNP to track the movement and interaction of radiation throughout the CAD [27]. Domain decomposition partitions of the spatial domain into subdomains that can fit the memory limitation. While domain decomposition Monte Carlo simulations can improve the efficiency of simulation run time on HPCs, particle parallelism has been more successful at improving performance.

4.2 Surrogate Modeling

Surrogate modeling has emerged as a powerful tool in engineering design and analysis, particularly for systems characterized by computationally expensive simulations or experiments. Surrogate models, also known as metamodels, can provide computationally cheap mathematical representations of computational models. They approximate the relationship between input variables and outputs using data from simulations or experiments. Common types of surrogate models include Gaussian process models (Kriging) [28–30], polynomial response surfaces [31], neural networks [32], and support vector regression [33].

The models are constructed using a design of experiments (DoE), which strategically selects data points to train the surrogate. The trained surrogate then provides computationally efficient predictions of system behavior, enabling faster design iterations and optimization. Data from different sources of varying fidelities can also be used to construct surrogates or ensembles of surrogates [34].

In multi-physics systems, surrogate models are particularly useful for coupling multiple interacting physical phenomena, such as fluid-thermal-structural interactions in hypersonic flight. These systems often involve high-fidelity computational fluid dynamics (CFD) and finite element analysis (FEA), which are computationally expensive. Surrogate models provide a computationally efficient alternative by approximating the outputs of these high-fidelity models.

For example, [35] demonstrated the use of Kriging surrogates of Reynolds-Averaged Navier–Stokes (RANS) CFD predictions to simulate experimental data for aerodynamic pressure and heat flux in hypersonic vehicles. These surrogates were integrated into a Bayesian network for model calibration and uncertainty quantification, enabling efficient exploration of the design space to design wind tunnel experiments.

Surrogate modeling offers several advantages in engineering design:

- **Reduction in Computational Cost**: By replacing expensive simulations with surrogates, the number of calls to high-fidelity models is significantly reduced.
- **Support for Optimization**: Surrogates provide approximations of objective functions and constraints, making it easier to locate global optima in complex design problems [36].
- **Flexibility in Data Integration**: Surrogates can incorporate data from various sources of different fidelities, including simulations and experiments, to improve predictive accuracy.
- **Facilitation of Uncertainty Quantification**: Surrogates can be integrated into Bayesian frameworks to quantify uncertainties in model predictions and calibrate parameters using experimental data (Villanueva and Smarslok 2017).

Although surrogates can aid in reducing computational costs, practitioners should take precautions when using surrogate models. Surrogate models should be trained on relevant data and with a large enough amount of training data to be sufficiently accurate. Poorly trained surrogates can lead to inaccurate predictions. Additionally, surrogate models generally have increased extrapolation error when used outside of the range of the training data. This is particularly true for nonlinear systems.

5 Considerations for Modeling and Simulation in Classified Environments

Modeling and simulation in classified environments requires proactive planning and careful execution to ensure compliance with security protocols, operational efficiency, and the protection of sensitive data. The following discussion addresses key considerations for M&S activities in such environments, focusing on software design, adherence to security requirements, data security and access control, secure communication and collaboration methods, potentially relevant restrictions on software and tools, and hardware and infrastructure challenges.

5.1 Designing Software to Leverage Approved Packages and Provisioned Environments

Developing software for classified environments often begins in unclassified settings, provided the algorithms and software are not classified. This approach allows broader collaboration and faster development cycles. However, developers must ensure that any libraries and dependencies used in the unclassified environment are already approved for use in the classified environment. If that is not the case, suitable alternatives should be identified, or the approval process for the required libraries initiated as early as possible. Delays or denials of approvals can significantly impact project timelines, and in some cases, necessitate rewriting software to remove unapproved dependencies. Developers must ensure that the library versions used in the unclassified environment match those available in the classified environment. While bug fix releases are typically easier to approve, minor and major updates may require lengthy approval processes. Containerization can streamline deployment and reduce dependency-related issues.

Separate development, staging, and production areas are recommended for software systems requiring an Authority to Operate (ATO). Developmental areas use representative data, while staging areas allow verification and validation with authentic data. Production areas handle real-world data and require compliance with ATO or continuous ATO (cATO) processes. The use of DevSecOps practices, including continuous integration and deployment pipelines, can facilitate software updates and ensure compliance with security requirements.

5.2 Adherence to Security Requirements Within the Defense Industry

Security requirements in classified environments are governed by frameworks such as the Risk Management Framework (RMF) [37] and DoD Instruction 8510.01. These frameworks outline the controls and processes necessary to obtain an ATO for systems processing DoD data. Developers must understand the inherited controls available in the target environment and leverage them effectively. For example, integrating with provided Identity, Credential, and Access Management (ICAM) systems can streamline user management and enhance security.

The choice of libraries and third-party tools must be deliberate, considering factors such as active support, maintenance, and origin. Developers may be required to justify the use of specific software, including its development and maintenance history, to gauge the risk of introducing it into the classified environment. Ensuring precise alignment of dependencies across development, staging, and production environments reduces unexpected issues and promotes seamless, coordinated updates.

5.3 Ensuring Data Security and Access Control

Data security and access control in classified environments apply to the M&S activities. Sensitive data must be handled within designated controlled areas, which feature strict access controls, soundproofing, and electromagnetic shielding. Access to high-side systems should be established as early as possible, as delays can hinder timely software development and creation of analytical output.

Personnel with security clearances must adhere to strict protocols for handling classified information. Training on classification levels, proper procedures, and risks of accidental disclosure is essential. Collaboration tools, such as those provided by the Defense Information Systems Agency (DISA), offer secure web conferencing, instant messaging, and file sharing, backed by strong authentication methods like the Common Access Card (CAC).

M&S activities often present challenges to the data classification management. The aggregation of individually unclassified or lower-classified data points can reveal higher-level classified information, necessitating proactive classification of outputs. Models and simulations must be developed, run, and analyzed within secure environments, with data transmitted over encrypted networks.

5.4 Secure Communication and Collaboration Methods

Secure communication and collaboration for M&S in classified environments are important but often complex. Defense-in-depth strategies combine physical and technical safeguards to protect national security information. Controlled spaces are available for classified discussions, while isolated systems with NSA-approved Commercial Solutions for Classified (CSfC) frameworks ensure secure communications. CSfC frameworks use “two independent tunnels” setups to protect classified data [38].

Collaboration tools designed for classified environments enable secure interactions among stakeholders. These tools incorporate strong authentication and access controls to minimize human errors, such as accidental disclosures. The ability to revoke access or unsend data further enhances security.

Secure collaboration is important throughout the lifecycle of models and simulations. From the initial concept to final results, strict security protocols must be maintained to prevent breaches. The fidelity and accuracy of models may themselves be classified, requiring protection alongside the data they generate. Secure platforms and encrypted networks ensure that sensitive insights derived from simulations remain protected.

5.5 Restrictions on Software and Tools

Some software and tools require that restrictions be in place, to limit user access. System administrators can achieve this through implementation of access controls, usage policies, and system configurations to ensure secure operation. System administrators can define role-based access control to grant users permission to access certain software, tools, or areas of the HPC based on their roles. Software environment management tools can also be used to restrict which versions of software are available to users or user groups. Job schedulers such as SLURM or PBS can be used to enforce limits on resource usage, such as CPU time, memory, or GPU access, to prevent misuse and ensure proper allocation of resources. System administrators can also leverage audit logging and enforcing user authentication to further enhance control. Together, these strategies help maintain system integrity, optimize resource utilization, and ensure compliance with the system owner and/or project specific requirements.

The use of export-controlled software on an HPC is one example where software must be closely regulated by the HPC administrator. MCNP is an example of a code that is export controlled under the jurisdiction of the U.S. Department of Energy, 10 CFR 810, or the U.S. Department of Commerce, 15 CFR 730–744 [39]. It is distributed by Oak Ridge National Laboratory Radiation Safety Information Computational Center (RSICC). RSICC further restricts the distribution of MCNP, by limiting distribution of the software under guidance issued by the U.S. Department of Energy's Office of Nonproliferation and Arms Control. If an HPC user would like to use MCNP on the system, it is the HPC administrator's responsibility to verify that the user has an MCNP individual user agreement and license. This ensures that the user has been cleared through export control and nonproliferation review.

5.6 *Hardware and Infrastructure Constraints*

The deployment and operation of an HPC system or systems comes with significant hardware and infrastructure constraints that can limit performance, scalability, and cost-effectiveness. Some challenges include power consumption, cooling requirements, limits in physical space available, and network infrastructure, all of which must be addressed to ensure the sustainable and efficient use of HPC resources. One of the most pressing constraints is energy consumption. HPC systems are composed of thousands of high-performance processors and GPUs that consume significant amounts of electricity. A single HPC cluster can require several megawatts of power, and as systems scale upwards, energy demands increase exponentially. As such, identifying sustainable and scalable energy solutions is critical for the efficient use of the HPC.

Cooling is another major concern. The heat generated by densely packed computing nodes must be efficiently dissipated to prevent hardware failure. Traditional air-cooling methods are insufficient for modern HPC systems, leading to the use of more advanced cooling techniques such as liquid cooling and immersion cooling. However, these methods, while effective, require specialized infrastructure and additional energy to operate, further compounding the power consumption issue.

Physical space and network infrastructure pose limitations to HPC scalability. HPC systems require extensive floor space, specialized cabling, and high speed interconnects to maintain low-latency communication between nodes. The construction and maintenance of these facilities involve significant capital investment and ongoing operational expenses. Storage systems must also transfer vast data throughput and capacity, adding to the complexity and infrastructure load. Nuclear energy is being explored as a viable power source for HPC facilities, due to its stable and continuous energy supply with lower carbon footprint. Small modular reactors (SMRs), in particular, are being investigated as a potential solution for powering data centers and HPC clusters due to offering a compact, safe, and efficient energy source that can be deployed near facilities. Exploring alternative energy sources like nuclear power, presents an opportunity to overcome the power and cooling needs as HPC systems continue to scale.

6 Conclusion

Modeling and simulation is an important tool for addressing the complexities of defense acquisitions and modern military operations. It enables defense organizations to analyze, prepare for, and navigate multifaceted scenarios in a controlled and cost-effective environment. Conceptual models serve as the cornerstone of effective M&S, simplifying the inherent complexity of defense systems into manageable representations. These models guide the selection of appropriate techniques and resources, ensuring computational feasibility, fostering stakeholder alignment, and enhancing the validity and utility of simulation studies.

Defense M&S faces unique challenges, including uncertainty quantification, computational cost reduction, and the integration of multi-domain perspectives. Addressing these challenges requires innovative approaches, such as probabilistic design methods, surrogate modeling, and leveraging high-performance computing (HPC) systems. These strategies optimize resource allocation, reduce over-design, and enable efficient exploration of design spaces.

In classified environments, M&S efforts must adhere to stringent security protocols to protect sensitive data and maintain operational integrity by ensuring data security, access control, and compliance with frameworks like the Risk Management Framework. Secure communication methods, software restrictions, and proactive planning for hardware and infrastructure constraints are essential for successful M&S deployment.

The integration of technical and security aspects is paramount for effective M&S frameworks in the defense industry. By aligning technical innovation with robust security measures, defense organizations can maximize the strategic advantages of M&S while safeguarding national security interests. Ultimately, M&S serves as a powerful enabler for military training, analysis, decision-making, and capability development, providing a strategic advantage in navigating the complexities of modern warfare.

References

1. National Training and Simulation Association, "A Primer on Modeling and Simulation," Arlington, VA, 2024.
2. R. Cloys, "Modeling and simulation—Exploring lethality before it's built," Air Force Materiel Command, 7 5 2025. [Online]. Available: https://www.afmc.af.mil/News/Article-Display/Article/4177397/modeling-and-simulation-exploring-lethality-before-its-built/. [Accessed 12 6 2025].
3. S. Robinson, "Conceptual Modelling for Simulation Part I: Definition and Requirements," *Journal of the Operational Research Society 59(3):,* vol. 59, no. 3, pp. 278–290, 2008.
4. J. J. Boomgaardt, V. Mojtahed and B. Waite, "NATO-MSG-058 Conceptual Modeling for M&S," NATO Science & Technology Organization, 2008.
5. S. Robinson, "Conceptual Modeling for Simulation," in Proceedings of the 2013 Winter Simulation Conference, 2013.
6. NASA, "M&S Definitions," [Online]. Available: https://public.ksc.nasa.gov/mns/ms-definitions/. [Accessed 2025].
7. S. Robinson, "Tutorial: Choosing What to Model—Conceptual Modeling for Simulation," in Proceedings of the 2012 Winter Simulation Conference, 2012.
8. Simulation Interoperability Standards Organization (SISO), "Simulation Conceptual Modeling (SCM) Study Group Final Report," SISO, 2006.
9. W. L. Oberkampf, S. M. DeLand, B. M. Rutherford, K. V. Diegert and K. F. Alvin, "Error and uncertainty in modeling and simulation," *Reliability Engineering & System Safety,* vol. 75, no. 3, pp. 333–357, 2002.
10. K. D. Rao, H. S. Kushwaha, A. K. Verma and A. Srividya, "Quantification of epistemic and aleatory uncertainties in level-1 probabilistic safety assessment studies," Reliability Engineering & System Safety, vol. 92, no. 7, pp. 947–956, 2007.
11. D. C. Villanueva, R. Hickey, A. Markina-Khusid and A. Chaudhuri, "Capturing Uncertainty in the Digital Twin State–Addressing Data Storage and Data Flow Bottlenecks," in AIAA SCITECH 2025 Forum, 2025.
12. M. C. Kennedy and A. O'Hagan, "Bayesian calibration of computer models," *Journal of the Royal Statistical Society: Series B (Statistical Methodology),* vol. 63, no. 3, pp. 425–464, 2001.
13. D. Higdon, M. Kennedy, J. C. Cavendish, J. A. Cafeo and R. D. Ryne, "Combining field data and computer simulations for calibration and prediction," *SIAM Journal on Scientific Computing,* vol. 26, no. 2, pp. 448–466, 2004.
14. A. M. Stuart, "Inverse problems: a Bayesian perspective," Acta numerica, vol. 19, pp. 451–559, 2010.
15. L. Biegler, G. Biros and O. Ghattas, Large-scale inverse problems and quantification of uncertainty, vol. 712, M. Heinkenschloss, D. Keyes, B. Mallick, L. Tenorio, B. van Bloemen Waanders, K. Willcox and Y. Marzouk, Eds., West Sussex, United Kingdom: John Wiley & Sons, 2010.
16. R. C. Smith, Uncertainty quantification: theory, implementation, and applications, Society for Industrial and Applied Mathematics, 2024.
17. D. M. Blei, A. Kucukelbir and J. D. McAuliffe, "Variational inference: A review for statisticians," Journal of the American statistical Association, vol. 112, no. 518, pp. 859–877, 2017.
18. Y. Marzouk, T. Moselhy, M. Parno and A. Spantini, "Sampling via Measure Transport: An Introduction," in Handbook of Uncertainty Quantification, Springer International Publishing, 2016, pp. 1–41.
19. N. Smith and S. Mahadevan, "Probabilistic methods for aerosapce system conceptual design," Journal of spacecraft and rockets, vol. 40, no. 3, pp. 411–418, 2003.
20. S. Kumar, R. T. Haftka and B. V. Sankar, "Probabilistic optimization of intergrated thermal protection system," in 12th AIAA/ISSMO Multidisciplinary Analysis and Optimization Conference, 2008.

21. D. Villanueva, R. T. Haftka and B. V. Sankar, “Accounting for future redesign to balance performance and development costs,” Reliability Engineering & System Safety, vol. 124, pp. 56–67, 2014.
22. F. Pukelsheim, Optimal design of experiments, Society for Industrial and Applied Mathematics, 2006.
23. J. P. Kleijnen, “Design of experiments: overview,” in Winter Simulation Conference, 2008.
24. F. A. Viana, “A tutorial on Latin hypercube design of experiments,” Quality and reliability engineering international, vol. 32, no. 5, pp. 1975–1985, 2016.
25. D. C. Villanueva, B. P. Smarslok and R. Perez, “Sequential experimental design and model calibration for targeted events,” in 18th AIAA Non-Deterministic Approaches Conference, 2016.
26. Los Alamos National Laboratory, MCNP Introductory Class Notes:, 2020.
27. B. B. Forrest and W. R. Martin, “HIgh Performance Computing and Monte Carlo,” in American Nuclear Society 2004 Winter Meeting, Washington, D.C., 2004.
28. J. P. Kleijnen, “Kriging metamodeling in simulation: A review,” European journal of operational research, vol. 192, no. 3, pp. 707–716, 2009.
29. T. W. Simpson, T. M. Maurery, J. J. Korte and F. Mistree, “Kriging models for global approximation in simulation-based multidisciplinary design optimization,” AIAA Journal, vol. 39, no. 12, pp. 2233–2241, 2001.
30. M. L. Stein, Interpolation of spatial data: some theory for kriging, Springer Science & Business Media, 1999.
31. R. H. Myers, D. C. Montgomery and C. M. Anderson-Cook, Response surface methodology: process and product optimization using designed experiments, John Wiley & Sons, 2016.
32. M. Smith, Neural networks for statistical modeling, Thompson Learning, 1993.
33. A. J. Smola and B. Scholkopf, “A tutorial on support vector regression,” Statistics and computing, vol. 14, pp. 199–222, 2004.
34. B. Peherstorfer, K. Willcox and M. Gunzburger, “Survey of multifidelity methods in uncertainty propagation, inference, and optimization,” SIAM Review, vol. 60, no. 3, pp. 550–591, 2018.
35. D. C. Villanueva and B. P. Smarslok, "Using expected information gain to design aerothermal model calibration experiments," in 17th AIAA Non-Deterministic Approaches Conference, 2015.
36. R. T. Haftka, D. Villanueva and A. Chaudhuri, "Parallel surrogate-assisted global optimization with expensive functions–a survey," Structural and Multidisciplinary Optimization, vol. 54, pp. 3–13, 2016.
37. National Institute for Standards and Technology (NIST), Risk Management Framework for Information Systems and Organizations: A System Life Cycle Approach for Security and Privacy NIST SP 800–37 Rev. 2, NIST, 2018.
38. U. N. S. A. (NSA), “Commercial Solutions for Classified (CSfC),” [Online]. Available: https://www.nsa.gov/resources/Commercial-Solutions-for-Classified-Program/. [Accessed 2025].
39. RSICC, RSICC System Administrator Guidance, 2021.

Harnessing Data for Defense: Challenges and Solutions

Jenny Aleman

Abstract The Department of Defense (DoD) Data Strategy was released in 2020. The first computer was created 75 years before that, temporarily "storing" about 200 bytes of data. As of 2025, we create, capture, copy, and consume almost 181 zettabytes of data daily. As technology continues to revolutionize data collection, storage, and analysis, the capability to perform big data analytics is critical for enhancing decision-making and operational efficiencies within defense. The important details of data identification, access/acquisition, and storage are often overlooked when organizations decide to employ data for analytic and machine learning applications. This chapter will walk through a hierarchy of defense data needs, in order to reach data actualization for analysis to support the warfighter. The DoD poses unique challenges in the sensitive nature of defense-related data, including privacy concerns, interoperability issues, regulatory compliance, and protection for data critical to national security. The chapter concludes with a call to action for both defense organizations and their counterparts in the private sector to enhance overall readiness through data actualization.

1 Introduction

As technology continues to revolutionize data collection, storage, and processing, the capability to perform data analysis on demand is critical for enhancing decision-making and operational efficiencies within defense.[1] Really leveraging modern tools and software to predict and suggest courses of action without human initiation or intervention is achievable but requires a baseline set of tools and steps. The important details of data identification, data acquisition, and storage are often overlooked when organizations decide to pursue a data strategy or implement advanced data tools.

[1] The views expressed are my own and do not necessarily represent the views of the United States Government.

J. Aleman (✉)
Washington, D.C., USA
e-mail: jennyaleman333@gmail.com

A. Weissman (ed.), *Women in Advanced Computing and Software for Defense*, Women in Engineering and Science, https://doi.org/10.1007/978-3-032-17466-6_5

Furthermore, the data for defense operations poses unique challenges due to the sensitive nature of defense-related data, including privacy concerns, interoperability issues, regulatory compliance, and protection of data critical to national security. Practical and optimized use of defense data requires a foundation of data availability and storage before pursuing data analytics and machine learning to support the warfighter.

There are many challenges to achieving data actualization for defense, but this chapter will focus on the aspects that can be addressed by government agencies internally and their industry counterparts. That is, individual government agencies and military services cannot wait for an executive order or centralizing agent to change regulations at the federal level before modernizing information operations, and industry and the private sector must do their part to ensure that they are creating shareable, transferable technology. Federal pay and workforce benefit constraints will continue to hamper the government's ability to be the sole inventors and innovators of defense technology. And, while there are research and innovation arms such as the Defense Advanced Research Projects Agency (DARPA), the funding dedicated to research projects for defense data applications pales in comparison to private ventures. It makes much more sense for the government to leverage commercially available technologies and apply and configure these innovations to defense needs.

2 The Defense Dataset

It is imperative to define and provide scope around the dataset. For this framework, defense data shall be defined as any information, in any format, gathered and managed by government agencies and defense organizations: (1) for the purposes of staffing, training, and equipping military personnel—in the United States, functions of the Military Services and (2) for planning and executing operational missions, such as those used by the U.S. Combatant Commands. All responsibilities within any defense organization and its subcomponents will have both the standard business departments such as finance, human resources, legal, government relations, logistics, training, and information technology (IT), as well as defense-specific groups to support operational security and operations—whatever that operation may be within the scope of a nation's defensive and offensive military operations, whether it is weapon system maintenance or intelligence operations. This chapter will refer to the entire set of data for one defense organization as the defense system, which will include data pertaining to any responsibilities within a defense organization, whether it is directly defense-related or tied to business operations for the defense organization. Utilization of the framework and models discussed in this chapter are based on experiences within subgroups of the U.S. defense organization, however could be applied to other defense applications in other nation-states.

It is also relevant to note here that in 2018, the U.S. Government released a Federal Data Strategy [1], and many U.S. Federal agencies have followed suit in releasing guidance and policy, to include the U.S. Department of Defense (DoD). The DoD recognizes and encourages the Department to treat data as an asset, and the 2020 Data Strategy [2] outlines seven goals, shortened as VAULTIS, to become a data-centric defense enterprise and make sure that the data is visible, accessible, understandable, linked, trustworthy, interoperable, and secure. The strategies and overarching policies are constantly changing, especially considering the rapid change in software and applications for data use, and the increased inclusion of artificial intelligence (AI) and large language models.

3 Hierarchy of Defense Data Needs

In order to understand how to leverage data to its fullest extent, a familiar structure can be used to demonstrate the foundational layers needed. Maslow's hierarchy of needs is often referenced to organize human needs into a clear progression toward "self-actualization" and is depicted as a pyramid, with the lower level needs acting as prerequisites to achieving higher level needs. The structure demonstrates that it is difficult for an individual person to achieve their full creative and mental potential, if one does not have the foundational physiological, safety, social and belonging, and esteem needs. The DoD Data, Analytics, and Artificial Intelligence Adoption Strategy of 2023 [12] highlights the DoD Hierarchy of Needs as a progression from data to analytics to AI with governance and assurance as layers between. In this paper, Maslow's hierarchy will be used to depict how to successfully harness data for defense at its most advanced levels with the minimum requirements needed. Consider the modified hierarchy in Fig. 1 for defining the methodology to achieve success for specific defense application use cases within a larger system of defense data. That is, this framework is to be used at a lower level of operations within a defense organization, rather than the entire system of defense operations. Most stakeholders are concerned with how to leverage data for their immediate use and specific use case, so this is the most practical level at which an organization can operate. This framework may not be applicable to organizations as a whole, so it should not be applied systemically. There are also some assumptions that the organization has base level data operations, tools, and skill sets available for leveraging in order.

3.1 Data Knowledge Areas Not Applicable for Specific Use

In the context of the Data Management Association (DAMA) wheel [3], Fig. 1 for specific use cases ignores the three data management knowledge areas of Data Governance, Document and Content Management, and Data Warehousing and Business Intelligence, since these consider the entire system of data.

Regardless of whether an organization has Data Governance in place, this model assumes that the base of the pyramid is either leveraging a governance structure to obtain the data or acquiring the data without a defined process or management

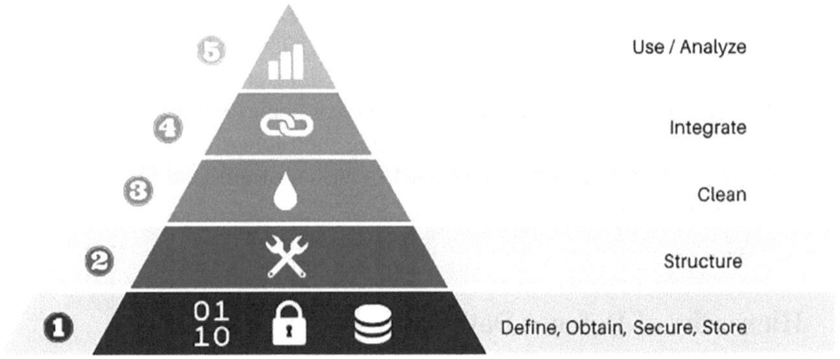

Fig. 1 Defense data needs for a use case, as adapted from Maslow's hierarchy

structure in place. The DMBOK highlights Peter Aiken's framework using DMBOK functional areas, which places data governance at the foundation of a pyramid, with organizations potentially progressing through phases that bounce from the middle to the bottom, before making it up to the top of the pyramid to use advanced data practices. Technically one can do effective data analysis without governance if you have an identified dataset—however, if governance is not taken into consideration, it will be difficult to integrate a final solution back into an overarching defense system.

Document and Content Management consists of the "capture, storage, access, and use of data and information stored outside relational databases" and ensures that an organization enables a relationship between structured and unstructured data [4]. The DAMA framework for the knowledge area of Document and Content Management suggests that an organization has to set up business rules to ensure standardization of the relationship between structured and unstructured data; in this defense data hierarchy, the dataset in scope, whether structured or unstructured, is isolated from the larger system. Unstructured data consists of "any document, file, graphic, image, text, report, form, video, or sound recording that has not been tagged or otherwise structured into rows and columns or records" [5] and the objective for a defense use case may include the analysis of unstructured data. However, the foundation in this model only requires that the data in scope is well defined relative to total data in scope for this use case, as opposed to the dataset at large within the defense system. While the ability for the dataset in scope and the analysis to comply with system standards is important for sustainment, the initial application may ignore the larger defense system in which it resides. This an unfortunate reality, but how most analyses are conducted within organizations, without regard to or integration back into a larger system of data. Ideally, each iteration of use of this model should be re-integrated into a defense system to increase the overall knowledge of data within the system. One approach for sustainment is to catalog and tag the analysis within a data management process, and the result of this model becomes part of the set of data defining the defense system, and, depending on the final format, it could itself be part of the unstructured data within the larger system.

The third DAMA knowledge area that this model does not touch is Data Warehousing and Business Intelligence. Similar to Content Management, this concept generally applies to a large system of data, with the end goal of integrating data to provide enterprise-level decisions. The defense data hierarchy in this chapter will use a subset of what would be contained in a data warehouse, for the purposes of a specific business intelligence use case within defense or to help answer questions for some defense problem. Like the Content Management area, ideally, the use of this model and the insights gained from this model would tie back to the larger defense system's data warehouse and business intelligence catalog. The three knowledge areas that this model does not touch on relate to requirements and recommendations for a larger system such as a system of defense data. The model helps stakeholders realize and utilize data within a defense system, without regarding how it fits back into a larger system.

Before beginning on the path to establish a robust pyramid base, the analysts and stakeholders must clearly define the defense use case and the specific problem that they are trying to solve. Oftentimes, people have a set of data and think creating a chart or simply summarizing the data translates to productive, optimized use of data for decision-making. That is, to effectively harness the power of data, one must begin at the bottom of the pyramid, with objectives and the appropriate dataset and tools, and an idea of what they want the tip of the pyramid to accomplish.

3.2 *Define, Obtain, Secure, and Store*

At the base (level 1) of the pyramid in this framework sits four different DMBOK data management knowledge areas, either in their comprehensive form, or a subset of information: Data Modeling and Design, Reference and Master Data, Metadata Management, and Data Storage and Operations.

The first step in getting to data actualization for a specific defense data scope is defining the data required to address the defense use case. It is not the systematic process of **Data Modeling and Design** for an entire organization, but the hyperfocus of identifying data elements needed within boundaries to make a comprehensive assessment of the use case in question. An example use case is when a defense organization wants to ensure traceability and maintenance of a specific drone weapon through its lifecycle of deployments. The data model will help establish common vocabulary for stakeholders through stakeholder engagements and documentation of subject matter expertise. Who are the people involved in the drone maintenance and deployments and do we want to capture this information? What are the different parts of the drone and to what level of detail do we want to trace their location and condition? When are there specific points of entry and exit throughout the processes of its deployment lifecycle that we can leverage to track status and/or location? Where is this data located currently, and where do we want it to go, if anywhere? Why are these events or transactions relevant or of interest for tracking the weapon system's lifecycle? How do we know when events or transactions related to the drone's status have occurred? How do we obtain the data for each of these milestones?

Still at the base of the pyramid of needs to properly analyze data for this use case is a subset of an organization's already-defined **Reference and Master Data**. When answering the questions to model the data, an analyst will likely run into inconsistencies in naming conventions or values of similar data elements between datasets required for the job. As an example, the system that tracks the maintenance operations of the drone may use the Original Equipment Manufacturer's part numbers to track parts that require repair, while the drone deployment system only refers to the drone with a deployment identifier. These inconsistencies between datasets describing the same features will require reconciliation before moving on. In a data-mature organization, the analyst could just refer to Master Data values previously defined for the datasets in scope. If the defense analysis use case is emergent, the analyst could ignore this requirement, and only define all data values within this use case, isolated from the rest of the defense system. Ignoring reference and master data is ill-advised, because it could invalidate results or be misconstrued if the data is taken out of context or re-integrated back into the system without the appropriate due diligence on its relevance and meaning. A basic example of this is the need to define and tie appropriate metadata and definitions to acronyms. During a meeting this week, one colleague was talking about the status of an Implementation Agreement, referring to it as an IA, between two organizations, and another colleague chimed in to state that they could follow-up on "that IA" with her cybersecurity team if needed, thinking that the first was referring to Information Assurance. Without the appropriate context, tagging, and reference to a standardized set of conventions to ensure you are talking about the same thing, the end result of an analysis could be misused or drive poor decision-making.

Similarly, the answers to the questions that define the Data Model will allow a multitude of **metadata** to surface: information about the processes and existing (or non-existing) data structures. In the drone traceability use case, the location of each dataset identified will have associated metadata about system information, update frequency, and data ownership. However, one may still need to identify any data rules and constraints of the data through experts of the datasets. As an example, the analyst may find out that when a repair is done to the drone, the operator at the depot has to manually input the repair status when they complete the repair; this results in a very error-prone set of records, and you may want to ensure that your Data Model tags that value as manual and needing additional processing support in the future. This will play into later requirements of data quality, but at a minimum, the potential gap should be identified at the base level.

Understanding the data is distinct from actually obtaining or being able to ingest or control the data; obtaining the data is still at the core of what's needed. To effectively obtain the data for use, the analyst must identify the data required and understand the basics of the data for ingestion. Note, just gaining access to a dataset will likely not meet the need, unless the data is already in an environment with the tools needed for analysis. Obtaining the dataset may mean setting up a connection or several connections with databases or systems with the source data which could be owned by other stakeholders. Obtaining drone data in the example use case for tracking the lifecycle of a drone may include porting in data from original equipment manufacturers, data from drone depot repair sites, as well as system data downloaded from the drone after each of its deployments.

A large part of data challenges usually comes before the data identification stage: the requirement owners have a problem set, but do not know what data would help, where to get the data, or even how to get the data to gain valuable insights. There can be a multitude of reasons why this might be difficult, but that won't be the focus of this chapter since it can really vary by situation and organization. In large organizations, and in defense organizations that are part of a nation's government, there are simply too many data owners and systems, so getting data to be collated in one place is a challenge. **Obtaining** the appropriate, complete, and accurate dataset is a large portion in the basic needs of analyzing data for defense.

Unfortunately many defense organizations do not technically progress past this stage of data needs, as the act of obtaining the data is arduous or meticulous. Barriers can be anything from lack of data rights (owned by industry), inability to identify authoritative source and "truth" of data, or simply that the data does not currently exist. Organizations within the defense community such as the combatant commands "remain heavily reliant on hard-copy reports, PowerPoint briefings, emails, and phone calls for critical updates" [6]. The unstructured data is not an issue in itself, it is the identification and acquisition of relevant data, whether structured or unstructured, that hinders organizational progress in achieving data actualization for even miniscule use cases, let alone for an enterprise system of defense data.

In a hierarchy of data needs not for defense, the DMBOK knowledge area of **Data Security** could reside in the second level, where Maslow's original concept identified Safety and Personal Security Needs. While humans technically do not need the feeling of security and safety to survive, the concept of data as an asset in defense deems data security a basic requirement. That is, in defense, if critical data is not secured, it may as well not exist, because otherwise it would jeopardize the security of a nation. In the context of defense operations, it is imperative and a base requirement that the data be securely stored and managed; otherwise, the defense mission is at risk. If an individual or unit is looking to acquire data and perform analysis and they do not consider the security of the data, they should not perform the analysis—they may as well be committing treason. Before acquiring the data, the analyst must define the conditions and access controls required to appropriately transfer, store, and share the data. The types of data within a defense data system requiring additional protections and controls could consist of:

- personally identifiable information (PII), a non-defense-specific set of data that could disclose personal information and could pose a threat or safety risk to an individual;

- personal health information (PHI), while not specifically for defense, PHI could be present within a defense system, if the defense system generates or stores data about military service members' health for readiness purposes;
- information at different classification levels, defined by the nation-state with security classification guides, or similar, to categorize data into different levels of severity of damage if the information is disclosed;
- sovereign data, which is generated and needs to be managed according to the laws, regulation, and policies specific to its originating country or region (this is relevant in defense partnerships and foreign military sales contexts); and, lastly,
- proprietary information or intellectual property, relevant in cases where the defense system leverages the defense industrial base for development and sustainment of technologies, and especially in peacetime, these rights are to be legally protected.

The methods for ensuring data security are not only deploying tools that have embedded security controls, but also include implementation of role or attribute-based access controls. Within the DoD, the security of systems is further regulated, requiring strict implementation of a Risk Management Framework that ultimately results in an Authority to Operate within a network. The U.S. government also has a Federal Risk and Authorization Management Program that provides a "standardized approach to security assessment, authorization, and continuous monitoring for cloud products and services" [7]. While many of these processes have clear checklists, it is ultimately up to the individual defense organization and their authorizing official to accept the risk (or not) of deploying a technology within their environment. In this hierarchy of defense data needs—in order to execute a use case most efficiently, the analysts would do best to select scripts/libraries, applications, and environments that are already vetted, on networks that are already existing at the appropriate classification level. This is of course assuming that the defense organization for which the analysis is required already has a baseline set of authorized tools to transfer, store, and process data. Data Security is a clear basic need, but there are also a lot of readily available and secure software and tools for use in the DoD. In the case of tracking a drone weapon system's lifecycle, data security means ensuring that the drone itself has secure software to protect its data, and there is a secure mechanism to transfer the data to the depots; it also means that the environment where the final data is housed must be authorized and secure, and that the software used to process the data is also secure.

Finally in the very first level of needs to process defense data is a place to put the information. The **Data Storage and Operations** knowledge DMBOK management area covers this—and in a standalone use case for analytics, the storage doesn't have to cover the entire lifecycle of the data, but just the time period for when the data needs to be processed and used. As mentioned in the data security requirement, the most efficient way to progress up through hierarchy is to leverage established (and already secure and vetted) data storage mechanisms. Data storage considerations should include network/server, environment, containerization requirements, and database structure. The storage and operations requirements will depend on the type of data needed to support the use case (structured versus unstructured), how large the dataset is and factors like how often it gets updated or accessed. For the drone lifecycle use case, it is likely structured data that requires some regular downloads of information from maintenance depots and post-mission operations; a relational database could be employed to perform predictive maintenance recommendations. If the use case is mission guidance during a drone deployment, additional security, transmission, and analysis tools must be considered. For example, if the drone takes photo and video, the storage required is not only substantially larger, but the tools to analyze and support a mission real time are complex and likely require custom software. Storage requirements are highly dependent on the specific use case.

The base of the pyramid is the widest and contains the most difficult requirements to help ensure that you can leverage your defense data for a use case. Defining and obtaining the defense data are generally the most difficult for analysts, followed by ensuring the environment and tools are secure. Deciding on a storage mechanism should be relatively straightforward, but this might change based on what architecture is needed to tackle the problem set. As with Maslow's hierarchy, these four data requirements are like the physiological needs of a human being: without a clear set of data contained securely within an environment, one cannot progress to the next level of data actualization.

3.3 Structure

Once you have defined the defense dataset, obtained it, and stored it (or established data feeds into) in a secure environment, you proceed to the next level of the hierarchy of data needs and structure the data. According to the DMBOK, **architecture** is an art and science that helps "optimize the function, performance, feasibility, cost, and aesthetics" [8] and, in this case, the analyst must determine what architecture is needed at the project or use case level. The architecture or structure surrounding the required defense dataset will optimize the ability for stakeholders to gain value from the dataset. Since this paper is only discussing a narrow scope and not addressing an entire organization or system's structure, the analysis is best suited to leverage an existing architecture, and/or set up an enclave within a system with minimal infrastructure.

Some questions that can help drive the understanding of what structure or architecture is needed for the use case:

- **Data Source**: Where is the data coming from or where does it reside? How often is it refreshed or accessed? In the most basic application, the data will be static or refreshed within the same repository. For advanced problem sets, the data may have multi-source bi-directional recurring refreshes and need connections to or pipelines from other systems. In the DoD, a complex data source setup will likely require documentation and body of evidence for an Authority to Connect between two systems. Connections with a commercial system is less likely in defense, so commercial data is generally ported or replicated within the firewall.

- **Stakeholder/User Base**: Who are the different stakeholders needing access, and do they need certain abilities to edit the data? Do certain stakeholders need to be restricted from viewing certain population of the data?
- **Data Integration**: What are the relationships between the data? If there are multiple datasets, how do they need to integrate and are there dependencies between the data that need to be defined?
- **Data Processing**: What type of processing will need to be done to the data to make it work for the use case? If the analysis can be done with an existing commercial data visualization or analytics tool, less needs to be done to organize and structure the data.

In the second level of the hierarchy, the analyst needs to set up the structure of the data based on where the data is coming from, in anticipation of who will use the data and how it will be used. Some of this information may be discovered while defining and obtaining the dataset, but this level is what drives what the data tech stack looks like.

3.4 Clean

Once the analyst has defined, obtained, secured, and stored the data within a specific structure, they may consider the quality of the data in level three. Quality and clean data in this case refers to the degree or extent to which a set of data meets the end users' needs, that is, the value of the data is of the utmost importance. Through the process of defining and obtaining the dataset, the level of scrutiny and process should lead to a usable set of information for the defense use case. In this level of the hierarchy then, the analyst needs to assess which **Data Quality** values are most critical to the success of the use case analysis. In 2020, DAMA NL researches Dimensions of Data Quality to establish a standard for data quality through an assessment of commonality [9]. The most common data values that can be considered by an analyst include accuracy, completeness, consistency, currency, and validity. Taking the use case of tracking a drone through its lifecycle, the data values of currency may not be necessary since it is a longer-term use case.

3.5 Integrate

As mentioned in the DMBOK, **Data Integration and Interoperability** is dependent on many other areas of data management, and it is difficult to truly consolidate data for use without the first three base layers of the data hierarchy. If the data model, metadata, and architecture of the data are clean, and the data quality is trustworthy, the defense data should be good enough to integrate with other information. Referring back to the drone lifecycle use case, at this stage, the data may be ready to integrate with other weapon systems, or perhaps the maintenance data combined with the unstructured mission data to provide additional clarity operational impacts to overall longevity of the drone. In Maslow's original hierarchy, level four consisted of elements that made up "esteem"—for defense data needs, this is the level where the data is truly trustworthy and shareable, and if the data is adhering to an overarching data governance structure or data model, this data can be combined with other data to solve other problems in the defense organization.

3.6 Use or Analyze

The final and top level of the hierarchy of defense data needs, where the data is the best it can be, is the point at which the data will be best used to solve defense data problems. The use of a pyramid provides a noteworthy nod to the importance and weight of establishing a solid foundational dataset prior to employing the latest tool on it. Oftentimes–not only in defense organizations–staff are eager to deploy the latest technology and do not consider the importance of having a solid foundation of data to be able to capture insights using machine learning or artificial intelligence tools. Staff who rush to use these applications are also the same ones who do not recognize the presence or lack of security measures in place to securely execute the tools. True data actualization occurs only after meticulous work to define and understand the data. And yet, organizations everywhere purchase data visualization tools or AI tools that promise to derive meaning and insights for the organization. We must do better.

4 Selected Challenges and Realistic Recommendations for Defense Organizations

While this chapter mainly touches on the basics needed to successfully analyze data for defense, there are some critical items that every defense organization should focus on to further enable data-driven decision-making. There are four weaknesses I continue to run into within defense organizations:

1. the lack of visibility or access to data to solve problem sets;
2. inability to implement or use technology/software due to fear of cybersecurity risks;
3. lack of resources or time to keep up with technologies; and
4. once identified, getting stuck in the quagmire of formal acquisition processes.

Implementing the following recommendations enables the defense organizations to invest and implement the appropriate tech stack to harness data for defense.

4.1 *Enforce Due Diligence on the Data Baseline*

The U.S. DoD Data Strategy was released in 2020. The first computer was created 75 years before that, temporarily "storing" about 200 bytes of data. It is estimated that in 2025 globally, we create, capture, copy, and consume almost 181 zettabytes of data [10]—that is—181,000,000,000,000,000,000,000 bytes. The state of data growth in general is alarming, and data within defense enterprises is similar. The majority of the chapter discusses the sheer rigor needed to get a dataset to a valuable place in its lifecycle even for a particular use case. While cataloging all data in a defense organization is a monumental endeavor that will almost certainly be overshadowed by mission and business priorities, any organization can incrementally approach the task by cataloging data as use cases arise. Since a 2021 DoD memo on Creating Data Advantage [11], the organization has made progress in moving toward the declared VAULTIS goals. The DoD has since changed from the assertion of treating data as a strategic asset, to treating data as a product in the 2023 strategy [12]. The notion of treating data as a product is very apropos in the context of harnessing data for defense: and just as any defense organization shouldn't send faulty weapons into the battlespace, it would do good for organizations to treat data and associated tools similarly. For all new systems and requirements, implementing and following a data model or standard set of guidelines will do much to improve overall value of data.

4.2 *Accept Risk to Enable Modernization*

The security of datasets is incredibly important and a basic requirement for defense data as discussed. As such, the organizational structure and foundation of defense information systems are generally disaggregated systems, whether intentional or just antiquated. It is no wonder that government officials often experience uncertainty around the appropriateness of centralizing defense-related data. The fact is we must trust the Swiss cheese model in cybersecurity—through the combination of physical authorization, authorities to operate (ATO), identity management, the infamous DD Form 2875 System Authorization Access Request, application hardening, semantic layers, access controls, multi-factor authentications, and on and on. I have experienced defense organizations drowning in bureaucracy generated by the need for a comfort blanket of security: in order to advance, defense organizations must be more agile in modernizing. As of the writing of this chapter, the DoD Chief Digital and Artificial Intelligence Officer (CDAO) testified before Congress and vowed to streamline the ATO process [13]. In addition to this, authorizing officials must be willing to take risks so that the organization can flourish with new capabilities. In order to motivate risk acceptance, defense organizations must solidify and enforce processes to ensure constant, smart monitoring of data flows and potential vulnerabilities in new and existing applications. Without a more streamlined authorization process, it will continue to be difficult for the DoD to efficiently partner with the private sector to leverage the latest and greatest technology to optimize operations at each of the levels of defense data needs.

4.3 Keep up with Data Innovations

The rapid growth of technology and solutions is overwhelming, and no organization charged with the job of planning and executing defense operations will be able to quell the zealous nature of industry interest and support. It should be the job of leadership, within a technology, data, or research branch of an organization, to ensure that the organization is well informed of the available technology. This centralizing or standardizing branch should adopt and plan for the entire data lifecycle. Oftentimes disparate groups within an organization focus on buzzwords like artificial intelligence to solve their problems, when the issue is more rooted in the bottom rung of the needs hierarchy like being able to control or access data fundamentally.

Additionally, as an individual contributor in a government organization, they should set aside time and resources to educate themselves on the latest innovations. The Legislative Branch of the United States Government has the Congressional Research Service within the Library of Congress, with government staff dedicated to providing policy and legal analysis to committees and Members of the House and Senate [14]. While it would be difficult for the executive branch to build such an organization to advise on execution, there are private sector resources that defense organizations can leverage to stay informed of critical technological innovations. Organizations like Gartner and Forrester do the research and hire subject matter experts to make objective recommendations. With so many options for every application to define, secure, store, structure, clean, integrate, and analyze data, it is imperative to leverage critical and relevant research.

4.4 Leverage Smart Contracting

At least in the U.S., the defense organization often has trouble procuring and awarding advanced technology rapidly to ensure fairness and open competition. Individual organizations will not be able to rewrite the Defense Federal Acquisition Regulation Supplement (DFARS),[2,3] but they can leverage authorities like regulations and special vehicles like Other Transaction Authorities [16] to execute rapid contracting for technological modernization. There are innovators in technology that the defense agencies can tap into to help execute all areas of defense data needs, including being able to obtain the data necessary. The CDAO team within the DoD created an acquisition ecosystem that allows groups within the DoD to work with and acquire cutting-edge technologies [17]. The DoD recognizes the importance of agile contracting and the need to leverage the private sector to accelerate mission readiness. In Dr. Douglas Matty's testimony before the 119th Congress, he specifically called out the importance of adopting the best tools that the commercial sector has to offer [13].

5 Challenges and Plea to the Defense Industrial Base and Cloud/Software Companies

On the other hand, there are actions and considerations that the defense industrial base and software companies can do to support other challenges that I've seen in U.S. defense organizations. The first challenge is tied to the acquisition process as well and could be supported by a technology companies taking more flexible stances in pricing and license structures. The second deals essentially with prime vendor lock—the defense organization cannot do basic analysis due to the inability to acquire the data. Understandably the private sector has to protect their bottom line, but operating with the partnership mentality will lead to increased complex analyses and elevated defense knowledge base overall.

5.1 Fair and Reasonable Pricing Models

The DoD budget request for Fiscal Year 2026 is around $961.6 billion [18]. While the defense budget top line for science and technology is about $20.3 billion [19], the true level of funding for IT systems and data applications is difficult to estimate, since each of the Military Services, components, and associated organizations can elect to invest in IT to support their operations. The most interesting part is that once the large weapon systems and hardware are accounted for, the IT budgets within the organizations are generally squeezed dry. Consider reasonable and tiered pricing models for software and services. If a technical solution is the best and latest, it will be of no use to a defense organization if it cannot be procured and thus utilized.

[2] Although an executive order was released on April 15, 2025 entitled "Restoring Common Sense to Federal Procurement" which aims to achieve exactly this [15].

[3] At the time of this writing, though, there is an ongoing effort to overhaul the Federal Acquisition Regulation by Executive Order on April 15, 2025, though, which will impact the DFARS.

Current DFARS regulation binds the Government to fair and open competition and/or stringent justifications for sole sourcing, which oftentimes results in large turnover of environments and software applications. Many software companies provide trial licenses or prototype periods, but reasonable pricing allows for more robust testing. Considering the sheer number of environments and applications required to execute a defense mission, it becomes incredibly difficult to buy the best in class for every application.

5.2 Intentional Use of Proprietary Markings and Data Rights

In working with industry and commercial sector services, I've observed a steadfast approach by the defense industrial base and commercial solutions and services companies to maximize intellectual property claim to documents, softwares, and data. It is of course the beauty and right of capitalism to be able to profit from what is yours. However, the industry slow-down of trying to maintain data rights continues to impede progress at the very basic level of data acquisition for defense applications. This also comes into play when working through problem sets and forcing the government to limit distribution of the data and thus limiting the ability to collaborate or integrate the data for wider use.

6 Conclusion

The bulk of the work to truly harness the power of defense data sits at the most basic level of data management. Defining and obtaining data prove difficult due to inherent security requirements of defense organizations, but are exacerbated by complex Federal acquisition regulation and commercial sector adherence to data rights. For a defense use case, a subset of the DMBOK can be leveraged to take an analyst from data to insight. However, without taking the bigger picture data governance and warehousing—strategic initiatives that should span a defense organization—the use case will not be easily integrated back into a useful state. The DoD should emphasize the basics of data, expedite authorizations through smart risk acceptance, stay in tune with cutting-edge technology, and leverage non-traditional contracting methods to increase the breadth of potential solutions. At the same time, industry and the commercial sector must play its part in enabling streamlined operations at the most basic levels of data operations.

References

1. Federal Data Strategy (2021) *2021 Action Plan*. Available at: https://apps.dtic.mil/sti/pdfs/AD1151674.pdf (Accessed: 12 September 2025).
2. Department of Defense (2020) DoD Data Strategy: Unleashing Data to Advance the National Defense Strategy. Available at: https://media.defense.gov/2020/Oct/08/2002514180/-1/-1/0/DOD-DATA-STRATEGY.PDF (Accessed: 12 September 2025).
3. DAMA International (2017) *DAMA-DMBOK: Data Management Body of Knowledge*. 2nd ed. Bradley Beach, NJ: Technics Publications, pp. 35–36.
4. DAMA International (2017) *DAMA-DMBOK: Data Management Body of Knowledge*. 2nd ed. Bradley Beach, NJ: Technics Publications, pp. 287.
5. DAMA International (2011) *The DAMA Dictionary of Data Management: Over 2,000 Terms Defined for IT and Business Professionals*. 2nd ed. Denville, NJ: Technics Publications, pp. 92.
6. Quinn, B., Sickler, B. & Wiltse, D. (2025) 'Modernizing Military Decision-Making: Transforming European Command', War on the Rocks, 24 April. Available at: https://warontherocks.com/2025/04/modernizing-military-decision-making-transforming-european-command/ (Accessed: 12 September 2025).
7. General Services Administration (GSA) (2025) FedRAMP. Available at: https://www.gsa.gov/technology/government-it-initiatives/fedramp (Accessed: 12 September 2025).
8. DAMA International (2017) *DAMA-DMBOK: Data Management Body of Knowledge*. 2nd ed. Bradley Beach, NJ: Technics Publications, pp. 97.
9. Black, A. and van Nederpelt, P. (2020) Dimensions of Data Quality (DDQ) Research Paper. Version 1.2, 3 September. Available at: https://www.dama-nl.org/wp-content/uploads/2020/09/DDQ-Dimensions-of-Data-Quality-Research-Paper-version-1.2-d.d.-3-Sept-2020.pdf (Accessed: 12 September 2025).
10. Statista (2025) *Worldwide data created*. Available at: https://www.statista.com/statistics/871513/worldwide-data-created/ (Accessed: 12 September 2025).
11. Deputy Secretary of Defense (2021) *Creating Data Advantage*. Available at: https://media.defense.gov/2021/May/10/2002638551/-1/-1/0/deputy-secretary-of-defense-memorandum.pdf (Accessed: 12 September 2025).
12. Department of Defense (2023) *2023 Data, Analytics, and Artificial Intelligence Adoption Strategy*. 2 November. Available at: https://media.defense.gov/2023/Nov/02/2003333300/-1/-1/1/DOD_DATA_ANALYTICS_AI_ADOPTION_STRATEGY.PDF (Accessed: 12 September 2025).
13. Matty, D. (2025) 'Information technology and artificial intelligence posture of the Department of Defense', testimony before the Subcommittee on Cyber, Information Technologies, and Innovation, Committee on Armed Services, United States House of Representatives, 8 May. Available at: https://armedservices.house.gov/uploadedfiles/5.8_matty_testimony.pdf (Accessed: 12 September 2025).
14. Congressional Research Service (CRS) (n.d.) *CRS Reports*. Available at: https://www.loc.gov/crsinfo/ (Accessed: 12 September 2025).
15. The White House (2025) *Restoring Common Sense to Federal Procurement*. Available at: https://www.whitehouse.gov/presidential-actions/2025/04/restoring-common-sense-to-federal-procurement/ (Accessed: 12 September 2025).
16. AcqNotes (2024) *Other Transaction Authority (OTA)*. Available at: https://acqnotes.com/acqnote/careerfields/other-transaction-authority-ota (Accessed: 12 September 2025).
17. Tradewinds AI (2025) *Tradewinds: The CDAO's Acquisition Ecosystem*. Available at: https://www.tradewindai.com/ (Accessed: 12 September 2025).

18. Department of Defense (2025) *Fiscal Year 2026 Budget Request Overview Book*. Office of the Under Secretary of Defense (Comptroller)/Chief Financial Officer, June. Available at: https://comptroller.defense.gov/Portals/45/Documents/defbudget/FY2026/FY2026_Budget_Request_Overview_Book.pdf (Accessed: 12 September 2025).
19. 20. DAMA International (2011) *The DAMA Dictionary of Data Management: Over 2,000 Terms Defined for IT and Business Professionals*. 2nd ed. Denville, NJ: Technics Publications, pp. 1–5.

The Artificial Intelligence Factory: Concept, Implementation, and Impacts

Protima Banerjee

Abstract The Artificial Intelligence Factory (AIF) is a structured framework that enables the definition, design, refinement, and deployment of AI technologies. It serves as both the logical and physical infrastructure supporting the entire AI life cycle—from project conceptualization and data collection to model development, selection, training, validation, optimization, and ultimately, deployment and ongoing monitoring of AI-enabled systems. Beyond its technical foundation, the AIF also emphasizes the human-centric aspects of computing. It empowers "citizen developers" to harness AI as a force multiplier. Additionally, it fosters collaboration across disciplines by standardizing vocabularies and toolsets, bridging the gap between project managers, domain specialists, data scientists, and machine learning engineers. In this chapter, we describe the technological foundations of the AIF and propose that its true power lies in its ability to shape an organizational culture where AI is not just implemented but thoughtfully integrated. By ensuring AI is deployed with precision, rigorously evaluated, and aligned with human needs, the AIF fosters an environment where AI enhances decision-making, reduces cognitive burden, and operates as a trusted assistant.

1 Introduction

Over the last decade, Artificial Intelligence (AI) has evolved into a transformative force. Its ability to process vast amounts of data at speed, recognize patterns, and support decision-making has driven advancements in healthcare, finance, logistics, transportation, agriculture, defense, and a multitude of other sectors [1]. However, production-grade AI systems require a structured, scalable development and deployment process to ensure efficiency, repeatability, and reliability—it is this need that gives rise to the concept of the AI Factory (AIF) [2].

P. Banerjee (✉)
ASRC Federal, Reston, VA, USA
e-mail: pbanerjee@asrcfederal.com

A. Weissman (ed.), *Women in Advanced Computing and Software for Defense*, Women in Engineering and Science, https://doi.org/10.1007/978-3-032-17466-6_6

The AIF is an integrated framework supporting the entire AI lifecycle, from ideation to deployment. Central to this is a well-defined infrastructure encompassing data pipelines, model training platforms, validation systems, and deployment frameworks [3]. By standardizing these elements, the AIF ensures AI and machine learning technologies are maintainable and able to adapt as organizational objectives change.

At its core, the AIF is a human-centric computing paradigm. While AI automates complex tasks, its full potential emerges as a complement to human expertise. The AIF fosters collaboration between data scientists, engineers, and domain experts, ensuring AI applications address real-world challenges in a quantifiable and demonstrable manner. Additionally, it empowers citizen developers (see Fig. 1)—professionals with domain expertise but limited AI knowledge—by providing accessible tools and frameworks, democratizing AI adoption.

Standardization, especially data standardization [4], is another cornerstone of the AIF. Traditional AI projects often suffer from fragmented workflows, inconsistent methodologies, and siloed development. By introducing common vocabularies, interoperable toolsets, and best practices, the AIF streamlines AI development, accelerates innovation, and enhances cross-functional collaboration. This structured approach reduces inefficiencies and improves the scalability of AI solutions across the enterprise.

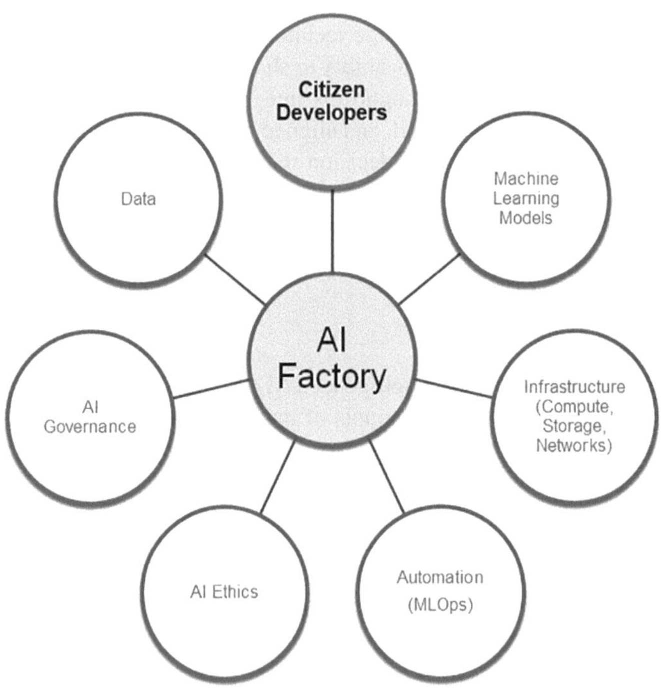

Fig. 1 AI factory is the intersection of AI technology and the citizen developer

Beyond infrastructure, the AIF represents a cultural shift in AI adoption. It promotes an AI-first mindset, embedding AI into strategic decision-making and operational workflows. This shift enables organizations to maximize AI's value, transforming industries through automation, efficiency, and innovation. Moreover, the AIF advocates for responsible AI, prioritizing transparency, fairness, and accountability. By aligning AI governance with ethical standards, organizations can mitigate biases, enhance trust, and ensure compliance with evolving regulations. The AIF is more than an operational framework—it is a strategic imperative for organizations seeking to harness AI effectively.

The remainder of this chapter delves deeper into the design, implementation, and impacts of the AIF. Section 2 examines the key stages of AI development within the AIF, detailing processes such as data acquisition, model training, validation, and deployment. Section 3 explores the human-centric aspects of AI development, including the role of citizen developers and the importance of interdisciplinary collaboration. Section 4 addresses the critical issue of sustainability, focusing on energy-efficient AI practices and responsible resource management. Finally, the chapter concludes with an analysis of the future of AIF, discussing emerging trends, challenges, and opportunities for organizations looking to integrate AI at scale.

2 Technical Framework of the AIF

The foundation of the AIF is a technical framework that ensures the seamless development, deployment, and monitoring of AI models. This framework consists of well-defined processes, standardized toolsets, and scalable infrastructure that collectively support an end-to-end development and deployment lifecycle. The technical framework of the AIF spans multiple layers, including data management, model training, validation, deployment, and observability, each playing a critical role in operationalizing AI at scale.

At the core of the AIF is the AI development lifecycle, which consists of eight essential stages: Defining the Problem, Collecting Data, Preprocessing Data, Model Training, Model Evaluation, Model Testing, Model Deployment, and Model Observability. Each of these stages is interconnected, with feedback loops allowing for iteration and continuous improvement. These feedback mechanisms ensure that AI models remain accurate, fair, and efficient as they are exposed to evolving data in the real-world. A "typical" AI development lifecycle is presented in Fig. 2.

Automation in the AI lifecycle is called MLOps. The term MLOps derives from DevOps, an established methodology that promotes automation, collaboration, and continuous integration and deployment (CI/CD) in the software engineering lifecycle. MLOps extends these principles to the AI and machine learning space by incorporating automation specific to data management, model training, model deployment, and model monitoring. While traditional DevOps primarily deals with code, MLOps must account for continuously evolving datasets and model retraining. MLOps pipelines include steps such as automated data preprocessing, experiment tracking,

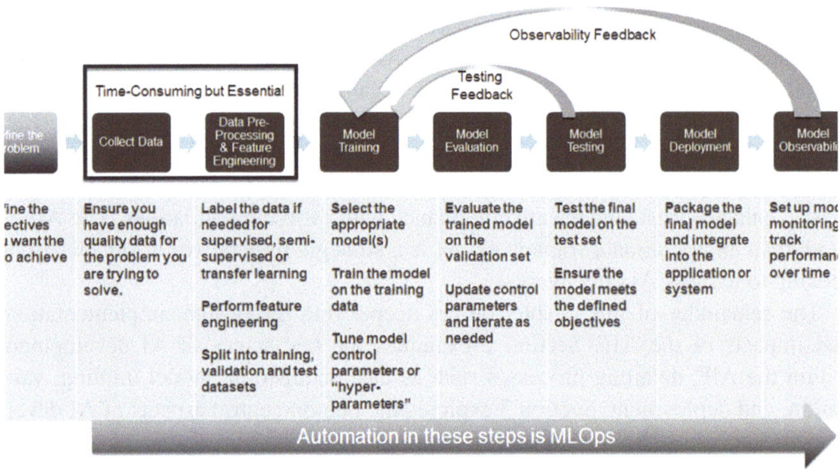

Fig. 2 AI development lifecycle

hyperparameter tuning, model versioning, and real-time model observability. The integration of feedback loops into MLOps ensures that deployed AI models maintain accuracy and relevance in an automated or semi-automated manner.

2.1 *Problem Definition and Data Collection*

The AIF process begins with problem definition, where teams identify the key objectives and business challenges AI will address. This requires collaboration between domain experts, project managers, and data scientists to ensure a clear alignment between AI solutions and strategic goals. A well-defined problem statement helps prevent misaligned AI initiatives and ensures that development efforts focus on practical, high-impact solutions. In many ways, this first step is the most critical of the AIF.

Having a well-defined problem statement is the most important step in the AI development lifecycle because it serves as the foundation for all subsequent stages. A vague or poorly scoped problem can lead to wasted resources, ineffective models, and misaligned AI implementations that fail to deliver value. A precise problem statement clearly articulates the business challenge, success criteria, and expected outcomes, ensuring that the AI team has a shared understanding of the objectives. It also helps in selecting the appropriate data sources, modeling approaches, and evaluation metrics. Without a clear problem definition, AI development becomes an exploratory exercise with no clear direction, ultimately leading to inefficiencies, increased costs, and suboptimal decision-making. A well-articulated problem statement ensures that

AI efforts are purpose-driven, measurable, and aligned with organizational goals, maximizing the impact of AI solutions.

Following problem definition, the next stage involves data collection. AI models rely on vast amounts of structured and unstructured data from various sources, including databases, IoT sensors, logs, and web scraping. The AIF provides standardized pipelines for ingesting, storing, and organizing this data to maintain consistency and quality. These pipelines incorporate data governance principles, ensuring compliance with privacy regulations and ethical AI practices.

A crucial aspect of data collection is data standardization, which ensures consistency across different data sources and formats. Standardized data enables integration, reduces errors, and improves the efficiency of AI model training. To standardize data at collection points, organizations should establish predefined schemas, enforce consistent naming conventions, and implement automated validation checks. These kinds of checks may already be present in Extract Transform and Load (ETL) pipelines that feed into data warehouses and data lakes.

Standardization efforts should also include normalizing data types, handling missing values in a systematic way across data sources, and applying transformation techniques to align diverse datasets. Additionally, metadata management should be integrated into the data pipeline to track lineage, provenance, and quality metrics, providing transparency and accountability in the data collection process. The AIF incorporates or leverages tools that automate these standardization steps, ensuring that raw data is preprocessed efficiently before it enters downstream AI workflows.

2.2 *Data Preprocessing and Feature Engineering*

Once data is collected, preprocessing and curation play a critical role in ensuring data quality. This involves cleaning, normalizing, and transforming raw data to remove inconsistencies and biases. The AIF incorporates automated tools to detect and handle missing values, anomalies, and imbalances, thus improving the reliability of AI models.

2.2.1 Feature Engineering

Feature engineering techniques further enhance data representation, optimizing the collected datasets for training machine learning algorithms. Feature engineering is a critical process in machine learning and AI model development that involves transforming raw data into meaningful "features" or characteristics that improve a model's ability to learn patterns and make accurate predictions. Feature engineering encompasses the creation or derivation of dataset features and selecting the most relevant ones to solve the problem at hand.

Features may be numerical, categorical, textual, or time-series data, each of which requires different transformation techniques. Techniques such as normalization, scaling, one-hot encoding, and embedding representations are commonly used to convert raw data into machine-readable inputs [5]. Additionally, feature extraction methods, such as Principal Component Analysis (PCA) are used to reduce dimensionality while retaining critical information [6]. For time-series and sequential data, lag features, rolling statistics, and frequency transformations can reveal important temporal patterns [7]. Similarly, natural language processing tasks often leverage techniques like tokenization, word embeddings (e.g., Word2Vec, BERT), and n-gram extraction to convert text into useful representations [8].

Feature engineering is iterative and involves extensive experimentation to identify the most informative variables. Automated feature selection techniques, including Recursive Feature Elimination (RFE), mutual information analysis, and Lasso regression, help eliminate redundant or irrelevant features, improving model efficiency. A well-engineered feature set significantly enhances model accuracy, reduces computational costs, and improves interpretability. Research in this domain has demonstrated that effective feature engineering can significantly outperform complex deep learning models trained on unrefined data [9].

2.2.2 Data Labeling

The data labeling process is a fundamental step in AI model development, particularly for supervised learning applications. It involves annotating raw data with relevant tags, classifications, or attributes, enabling AI models to learn meaningful patterns and relationships within the dataset. Accurate and high-quality labeled data directly impacts model performance, making labeling a crucial aspect of the AI framework. Various approaches to data labeling exist, each suited to different data types, project scales, and resource constraints:

- **Manual Labeling**: Human annotators review and label data points based on predefined guidelines. This approach is widely used for complex tasks such as image recognition, sentiment analysis, and medical diagnostics. However, manual labeling is time-consuming and expensive, particularly when large datasets are involved.
- **Programmatic Labeling**: Leverages rule-based heuristics, regular expressions, or weak supervision techniques to generate labels automatically. This method is particularly effective for structured data, such as transaction logs or sensor readings, where domain-specific rules can be applied to classify data points.
- **Semi-supervised Learning Model for Labeling**: Combines human and machine efforts by using a small set of manually labeled data to train an initial model, which then labels additional data autonomously. Human annotators review and correct model-generated labels, refining the dataset iteratively. This approach is highly scalable and balances efficiency with accuracy, making it useful for applications where fully manual labeling is impractical [10]. **Active Learning** techniques can

optimize data labeling by prioritizing the most uncertain or ambiguous data points for human review. By selecting only the most informative samples for manual annotation, active learning reduces the labeling effort while maximizing model performance [11].

Another labeling methodology currently in common use is **Crowdsourcing,** which distributes labeling tasks to a large group of contributors via online platforms. This method is cost-effective for large-scale labeling tasks, but it is difficult to ensure consistency across annotators and not appropriate for sensitive or proprietary datasets such as those that are typically used for mission-critical applications [12].

The AIF could potentially integrate some or all the above labeling strategies, automating as much of the process as possible while maintaining human oversight where necessary. By optimizing data labeling through intelligent workflows and robust validation techniques, the AIF ensures that AI models will be trained on accurate, high-quality data, leading to better performance during deployment.

We note here that data collection, feature extraction, data preprocessing, and data labeling are the most time-intensive parts of the AI workflow. Inserting automation into these steps is a critical capability of the AIF.

2.3 Model Training, Evaluation, and Testing

2.3.1 Model Selection

Model selection is a critical step in AI development, as the choice of algorithm determines model performance, efficiency, and applicability to the problem domain. The AIF provides a framework for selecting machine learning and deep learning models based on the complexity of the problem, data characteristics, and computational resources. Common approaches include linear models, decision trees, ensemble methods, and neural networks. The selection process involves benchmarking multiple candidate models using baseline metrics to determine the most effective approach.

2.3.2 Model Tuning and Hyperparameter Optimization

A **hyperparameter** is a configurable setting in a machine learning model that influences how the model learns patterns from data. Unlike model parameters, which are learned during training, hyperparameters are set before training begins and control aspects such as learning rate, regularization strength, the number of hidden layers in a neural network, or the number of trees in a random forest. Optimizing hyperparameters through techniques like grid search, random search, or Bayesian optimization can significantly impact a model's accuracy, efficiency, and generalization ability [13].

Once a model is selected, hyperparameter tuning optimizes its performance by adjusting parameters that control learning behavior. The AIF incorporates automated hyperparameter tuning frameworks that leverage parallel computing to efficiently search the parameter space. This ensures that models achieve peak performance while minimizing computational overhead.

2.3.3 Model Evaluation

Model evaluation is essential for assessing a model's predictive accuracy, fairness, and generalizability. The AIF supports various evaluation metrics that include quantitative assessments of model accuracy, precision, and recall, among other metrics. A comprehensive listing of the metrics used to evaluate machine learning and natural language processing algorithms is beyond the scope of this paper, but can be found in [14, 15]. Additionally, fairness and bias detection tools assess whether the model produces equitable predictions across different demographic groups, ensuring ethical AI deployment [16]. Evaluating model robustness under different conditions, including adversarial scenarios and data shifts, further enhances reliability.

2.4 Model Deployment

For practitioners in the defense industry, deploying AI models in mission-critical edge applications presents unique challenges and requirements that differ significantly from traditional cloud-based deployments. Edge environments demand real-time processing, low latency, and high resilience under unpredictable conditions. Unlike cloud-centric models that rely on centralized computational resources, edge AI models must function efficiently within resource-constrained devices while maintaining accuracy, robustness, and security [17].

Deploying AI models to edge environments requires specialized techniques to ensure efficient performance, low latency, and minimal resource consumption (see Fig. 3). Key considerations include the following:

- **Model Optimization**: Techniques such as quantization, pruning, and knowledge distillation reduce model size and computational complexity, enabling efficient execution on resource-constrained edge devices while maintaining acceptable accuracy.
- **Model Conversion**: AI models need to be converted into specialized formats like ONNX [18] to ensure compatibility with edge hardware and optimize runtime performance.
- **Hardware Acceleration**: Utilizing dedicated hardware such as GPUs, TPUs, or FPGAs enhances inference speed and energy efficiency, allowing AI models to run with minimal latency on edge devices.

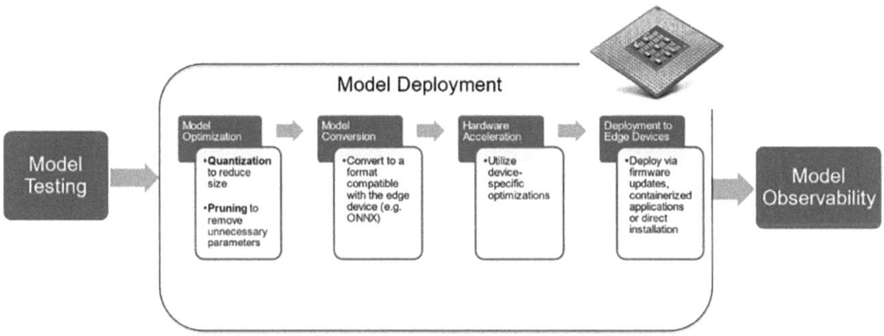

Fig. 3 Deployment to edge and internet of things (IoT) environments

- **Deployment to Edge Devices**: Optimized models must be integrated into embedded systems, IoT devices, or autonomous platforms, considering factors such as intermittent connectivity, real-time processing requirements, and power constraints.

To achieve seamless deployment to edge and mission-critical environments, the AIF must integrate robust DevOps and MLOps practices tailored to edge AI systems. Containerization technologies such as Docker and Kubernetes facilitate modular deployment across heterogeneous hardware platforms, ensuring consistency and scalability. Moreover, real-time monitoring and observability tools enable proactive model management by tracking inference latency, performance degradation, and environmental changes. By combining efficiency, security, and adaptive learning capabilities, mission-critical edge AI deployments can deliver reliable, autonomous decision-making while minimizing operational risks in high-stakes scenarios [19].

2.5 *Model Observability and the Feedback Loop*

Deployment strategies for AI models prioritize scalability, reliability, and integration with existing systems. Techniques like containerization using Docker and Kubernetes [20] ensure that models can operate consistently across different environments. Additionally, Continuous Integration and Delivery (CI/CD) pipelines facilitate seamless updates and minimize downtime.

Post-deployment monitoring is critical to identify issues such as data drift, performance degradation, or changing user needs. Tools like Prometheus and Grafana [21] enable real-time monitoring, while feedback loops facilitate iterative improvements. For instance, in fraud detection systems, ongoing monitoring ensures that models adapt to evolving threat patterns, maintaining their effectiveness. Observability and continuous monitoring are critical to sustaining AI models in production. The AIF employs model observability tools that track performance metrics, detect drift, and monitor fairness over time. These tools provide real-time insights into

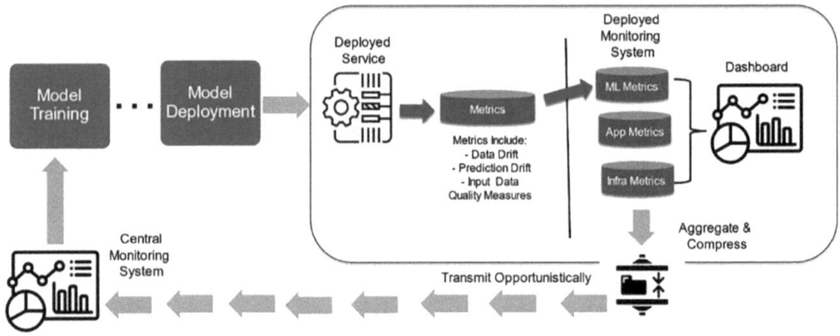

Fig. 4 Observability in environments with limited or intermittent connectivity

model behavior, allowing data scientists to address issues before they impact business outcomes. Feedback loops between observability and training ensure that AI systems remain adaptive and resilient to changing data conditions.

2.5.1 Monitoring Intermittently Connected Systems

In intermittently connected applications, such as those operating in remote military deployments or edge computing scenarios, model observability and continuous monitoring present unique challenges. Unlike traditional AI deployments where cloud-based monitoring ensures real-time tracking of performance, intermittently connected systems must function with limited or no network connectivity for extended periods. This necessitates a **hybrid approach to observability** in the AIF (see Fig. 4), where local monitoring mechanisms operate autonomously, and periodic data synchronization updates the centralized monitoring systems when connectivity is available.

One approach to ensuring observability in these environments is embedding lightweight monitoring agents within the deployed AI models. These agents collect critical performance metrics, such as inference latency, accuracy drift, and anomaly detection, storing the information locally until connectivity is restored. Techniques like federated learning and edge-driven logging help maintain the integrity of model performance tracking without constant cloud dependence. Additionally, decentralized monitoring architectures can facilitate peer-to-peer communication between edge devices, allowing them to share insights and adjust model parameters dynamically based on observed trends [22].

To address model drift in these constrained environments, organizations can implement adaptive learning strategies where models self-assess and adjust parameters based on local data distributions. When connectivity is re-established, batch synchronization of logs and performance data enables centralized retraining or fine-tuning of models, ensuring long-term reliability and consistency. By combining local observability with intermittent but structured cloud synchronization, AI models in mission-critical, intermittently connected applications can remain effective and resilient even in the face of unpredictable network conditions.

2.6 AIF Tooling

To implement the AI development and deployment processes, the AIF incorporates a comprehensive toolset spanning multiple layers of AI operations (see Fig. 5). At the data layer, tools for data ingestion, storage, preprocessing, and annotation ensure seamless data management. Model training tools include frameworks that enable efficient development of AI models. Testing and evaluation tools such as automated validation suites and adversarial testing platforms enhance model robustness. Deployment tools, including containerization technologies like Docker and Kubernetes, facilitate scalable AI deployment. Lastly, model observability tools provide monitoring, logging, and alerting capabilities to maintain AI integrity in production environments.

In this chapter, we specifically avoid calling out specific open source or licensed software products or hardware tools to implement the AIF. This is done for two reasons: (1) AI tooling changes at a remarkable pace and it is likely that any products called out here may be outdated or obsolete by the time of publication and (2) our goal is to remain vendor neutral and focus on the functions that the tooling provides.

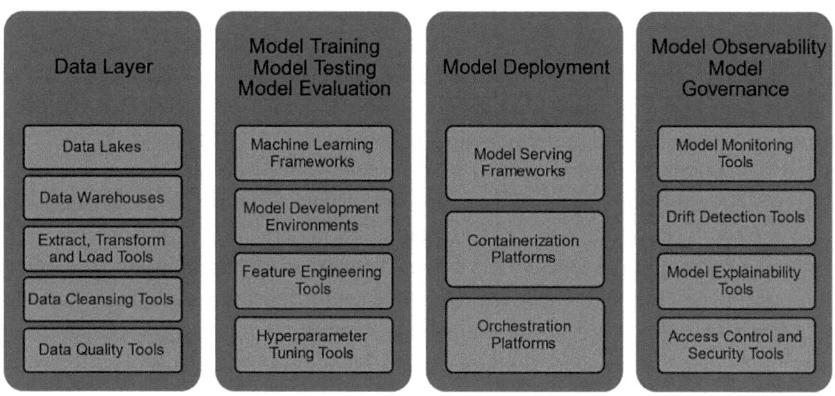

Fig. 5 AI Factory Toolsets

3 The Human-Centric Aspect of the AIF

The AIF is not solely a technical construct—its success depends equally on its ability to empower people, foster collaboration, and cultivate an ethical and democratized AI-driven culture. This section delves into the human-centric dimensions of the AIF, focusing on how it supports citizen developers, facilitates cross-functional collaboration, and embeds human-centric values into its design.

3.1 Enabling Citizen Developers

Citizen developers, defined as non-technical professionals who leverage AI tools to address domain-specific challenges, are a growing demographic within AI initiatives. Empowering these individuals to contribute meaningfully requires a thoughtful approach [23].

In the defense sector, citizen developers are military personnel, analysts, and operational specialists who leverage AI tools to enhance mission effectiveness without requiring deep expertise in data science or software engineering. These individuals could include intelligence officers who use AI-powered analytics to process reconnaissance data, logistics personnel who optimize supply chain operations through predictive maintenance models, and cybersecurity specialists who employ AI-driven threat detection systems (see Fig. 6). By providing user-friendly, low-code AI platforms and targeted training, the AIF enables these professionals to harness AI as a force multiplier, enhancing decision-making, operational efficiency, and situational awareness in dynamic and mission-critical functions.

Technical knowledge gaps limit the potential of citizen developers. The AIF shifts the focus from coding and implementation to problem-solving. In the AIF framework, citizen developers become a critical asset, bringing deep domain expertise to the table, which complements the technical skills of data scientists and engineers. By democratizing access the AIF ensures that solutions are rooted in practical, real-world insights that make sense to the citizen developers championing the project. A Navy commander might not need to understand the intricacies of neural networks but should be able to leverage AI-driven insights to optimize mission performance. By fostering an ecosystem where domain experts collaborate with AI tools effectively, the AIF expands the reach and impact of artificial intelligence and allows teams to develop clear success criteria for AI implementations.

3.2 Facilitating Cross-Functional Collaboration

The success of AI initiatives depends on effective collaboration across diverse teams, including domain experts, data scientists, software engineers, project managers, and

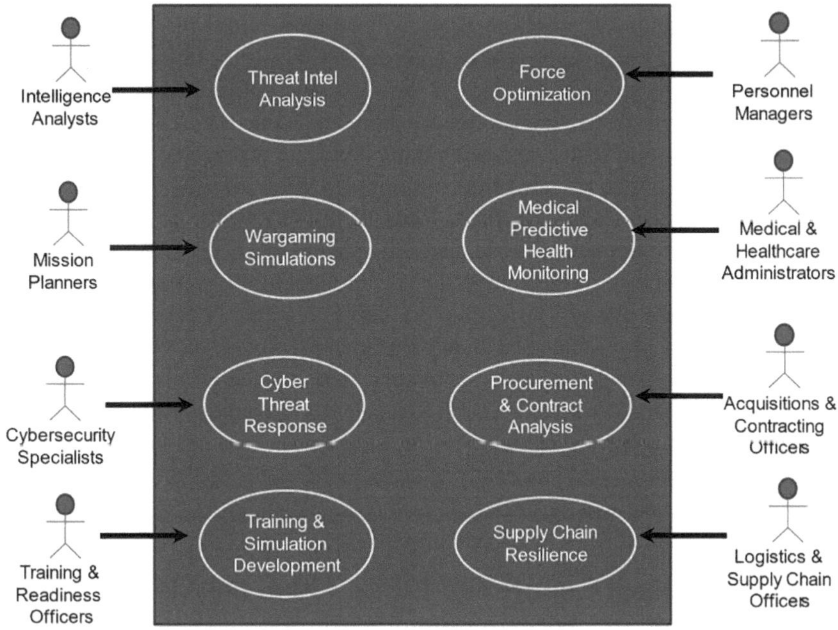

Fig. 6 Examples of AI factory citizen developers and their use cases

decision-makers. Each of these groups brings unique expertise to the table, but without a shared understanding, miscommunication can slow down progress and lead to inefficiencies. The AIF fosters cross-functional collaboration by not only creating structured workflows, but also by standardizing vocabularies to ensure all stakeholders operate with a common language.

Standardized vocabularies bridge the gap between technical and non-technical contributors, reducing ambiguity and aligning expectations. For example, a data scientist might use the term "model drift" to describe a degradation in AI performance over time, while an Army colonel may refer to it as "reduced accuracy" or "unreliable predictions." By formalizing definitions and ensuring all teams use the same terminology, misunderstandings are minimized, and problem-solving becomes more efficient. Organizations can achieve this standardization through internal glossaries, documentation, and AI governance frameworks that clearly define key concepts and best practices; these artifacts become part and parcel of the AIF.

Cross-functional collaboration in AI development also benefits from structured communication channels and shared tooling. When teams work with different tools or methodologies, integration challenges arise. The AIF promotes interoperability by establishing a common platform. This ensures that domain experts can easily interact with AI tools without requiring deep technical expertise while technical teams can efficiently translate domain knowledge into optimized AI solutions. It also keeps toolsets lean and technical vocabularies aligned with one another across functions.

A concrete example of cross-functional collaboration within the AIF can be seen in military logistics and supply chain management. Logistics officers must ensure that troops receive the right supplies at the right time, often under unpredictable conditions. AI can optimize this process by predicting demand, identifying the most efficient transportation routes, and automating inventory management. However, for AI to be effective, it requires input from multiple disciplines. Logistics officers define operational challenges and constraints, data scientists develop predictive models based on historical supply chain data, and software engineers integrate these models into existing logistics management systems. Throughout the process, standardized vocabularies ensure clarity—terms like "resupply interval," "demand fluctuation," and "critical stock levels" must be understood consistently across all teams.

By fostering collaboration and establishing standardized vocabularies, the AIF creates an environment where diverse teams can work seamlessly toward common objectives. This not only accelerates AI adoption, but also ensures that solutions are both technically robust and practically relevant to the challenges they aim to solve.

4 Organizational Culture and Ethics within the AIF

Organizational culture is the shared set of values, beliefs, norms, and practices that govern how employees behave within a company. In the context of AI factories, organizational culture is particularly important because the technology being developed has far-reaching societal impacts. This culture dictates not only how teams collaborate and communicate, but also how they approach challenges, innovation, and ethical decision-making. A **positive organizational culture** fosters an environment where creativity, critical thinking, and collaboration are encouraged. Leadership plays a crucial role in defining culture, setting the tone for how employees interact and how decisions are made. Communication, transparency, and mutual respect are essential components that contribute to a high-functioning work environment [24].

Continuous and successful AI development demands a culture of continuous learning and adaptation. The rapid pace of technological advancements and the inherent uncertainty of AI systems require an organizational culture that supports experimentation and iteration. AI teams need to feel empowered to take risks, test new ideas, and fail without fear of punitive consequences. This is crucial for driving innovation and ensuring that AI technologies evolve to meet the demands of users and society at large. Moreover, AI factories need **a culture of collaboration**, where multidisciplinary teams—comprising data scientists, software engineers, ethicists, and business leaders—work together toward shared goals.

This multidisciplinary approach can sometimes create tension as different fields prioritize distinct values—technical performance versus ethical integrity, for example. However, fostering a culture that values dialog and mutual respect allows AI teams to navigate these challenges more effectively, resulting in AI technologies that are both cutting-edge and ethically sound.

4.1 Ethics and the AIF

AI systems can have profound societal impacts, raising ethical issues that demand thoughtful consideration. One of the key challenges in AI development is addressing bias. AI models are trained on large datasets, and if these datasets are biased, the resulting models may perpetuate or even exacerbate social inequalities. For instance, biased data can lead to discriminatory outcomes in hiring algorithms or criminal justice prediction tools.

Fairness, accountability, transparency, and privacy are additional ethical considerations. AI systems often operate as "black boxes," with their decision-making processes hidden from users and even developers. This lack of transparency raises concerns about accountability, particularly when AI systems make decisions that affect people's lives—whether in hiring, healthcare, or law enforcement [25].

To address these ethical challenges, AI factories must establish clear ethical frameworks. These frameworks typically include principles such as fairness, accountability, transparency, and safety. For example, ensuring fairness may involve using unbiased datasets, conducting fairness audits, or developing algorithms that are explicitly designed to mitigate discrimination.

Ethics boards or committees within AI factories play a critical role in upholding these principles. These boards often consist of interdisciplinary teams that evaluate AI projects through ethical lenses, ensuring that designs adhere to established guidelines and remain aligned with social good. Furthermore, regulatory frameworks and external oversight may guide ethical practices, especially as governments and organizations push for clearer AI governance policies.

4.2 Balancing Innovation with Responsibility

AI development presents a constant tension between driving innovation and ensuring responsibility. AI factories are under pressure to produce breakthrough technologies, but with the potential consequences of such technologies, ethical considerations must always be top of mind. As companies race to deploy advanced AI models, they must balance the desire for progress with the potential risks of misuse or unintended harm.

Building an ethical organizational culture requires more than just developing ethical guidelines—it involves creating an environment where ethics are actively discussed, taught, and integrated into everyday practices. Training programs on ethics in AI, coupled with workshops, seminars, and open forums for discussion, ensure that all employees, from engineers to business leaders, are well-versed in the ethical challenges associated with AI development.

Ethical leadership is another key factor in fostering an ethical culture. Leaders who set a strong ethical example by making decisions that prioritize social good, transparency, and fairness help set the tone for the entire organization. These leaders encourage employees to consider the ethical implications of their work, thereby ensuring that ethical considerations are always factored into decision-making processes.

In AI factories, ethical considerations must be seamlessly incorporated into the daily workflows of AI development. This can be achieved by implementing regular ethics reviews throughout the development cycle—from conceptualization to deployment. Developers should be encouraged to consider the ethical implications of their design choices, and ethical audits should be a standard practice for all AI projects. AI companies can also create ethics decision trees to guide engineers in making morally sound decisions. These tools provide a structured way for AI teams to assess the ethical impact of their work, ensuring that they remain aligned with the company's broader ethical goals.

4.3 Encouraging Open Dialog

For a positive organizational culture to flourish, there must be an environment where open dialog is encouraged. Employees should feel comfortable raising ethical concerns and challenging the status quo when they believe something is amiss. This could mean flagging a potential bias in an algorithm, questioning a business decision that compromises transparency, or suggesting alternative methods to mitigate risks. Creating a culture where dissent is valued promotes integrity and accountability. Internal ethics committees or focus groups can facilitate these conversations, ensuring that concerns are addressed promptly and effectively.

5 Sustainability and the AIF

The AIF has a critical role to play in addressing the growing global emphasis on sustainability. As organizations increasingly adopt AI technologies, the need to integrate sustainable practices into the development and deployment of AI systems becomes essential [26]. This section explores the environmental dimensions of sustainability within the AIF, providing a comprehensive view of how AI can be a driver for sustainable innovation.

Environmental sustainability within the AIF focuses on energy efficiency, minimizing the carbon footprint, and optimizing resource utilization. AI models, particularly large-scale ones, demand significant computational power, leading to high energy consumption. To address this, the AIF must prioritize energy-efficient computing by implementing optimization algorithms, such as pruning and quantization, and leveraging renewable energy-powered data centers. Partnerships with cloud providers that champion green initiatives can enhance these efforts. The concept of "green AI"—minimizing the environmental impact of AI operations—further aligns the AIF with global sustainability goals. By actively monitoring energy usage metrics and incorporating lighter computational models, the AIF can balance innovation with environmental responsibility.

Minimizing the carbon footprint of AI projects is another critical objective. The AIF can achieve this by tracking and reporting carbon emissions associated with AI workflows, setting reduction targets, and employing strategies like geographic load balancing. For instance, routing workloads to data centers in regions with lower carbon emissions can reduce environmental impact. Conducting lifecycle assessments of AI systems allows organizations to quantify and mitigate their overall carbon footprint from development to deployment. Furthermore, recycling and reusing computational resources through practices like virtualization and distributed computing ensures maximum hardware utilization and reduces electronic waste.

Economic sustainability in the AIF emphasizes cost efficiency and scalability. The AIF can design modular AI architectures that dynamically scale based on demand, ensuring optimal resource allocation during both peak and low-usage periods. Serverless computing models, where organizations pay only for the compute resources they use, further enhance cost-effectiveness. Efficient resource management within the AIF is also achieved through DevOps and MLOps practices, which automate workflows, minimize errors, and reduce redundancies. Robust testing frameworks ensure that AI solutions are developed efficiently, saving time and resources.

AI itself plays a role in driving efficiency within the AIF. Predictive analytics tools can optimize resource allocation by identifying underutilized servers and redistributing workloads to improve productivity. AI-powered monitoring systems reduce downtime, saving energy and costs by maintaining smooth operational workflows. These approaches demonstrate how the AIF not only supports the development of AI systems, but also utilizes AI to optimize its internal processes, contributing to economic sustainability.

Building a culture of sustainability within the AIF requires leadership commitment and continuous improvement (see Fig. 7). Organizational leaders must set clear sustainability goals aligned with global frameworks. These goals should be supported by cross-functional teams tasked with innovating around sustainable AI practices. Measuring the impact of these initiatives is critical. The AIF can develop Key Performance Indicators (KPIs) for environmental, economic, and social sustainability, ensuring transparency and accountability through regular reporting.

Sustainability is an evolving objective, necessitating a culture of continuous learning and adaptation. The AIF should regularly review and update its sustainability practices, incorporating feedback from stakeholders to refine processes and align with broader goals. This iterative approach ensures that the AIF remains responsive to emerging challenges and opportunities in the sustainability landscape. By embedding sustainability into its core operations, the AIF not only minimizes its environmental impact but also drives long-term economic value. This holistic approach ensures that AI technologies contribute positively to local, regional, and global sustainability efforts while maintaining organizational resilience in a rapidly evolving landscape.

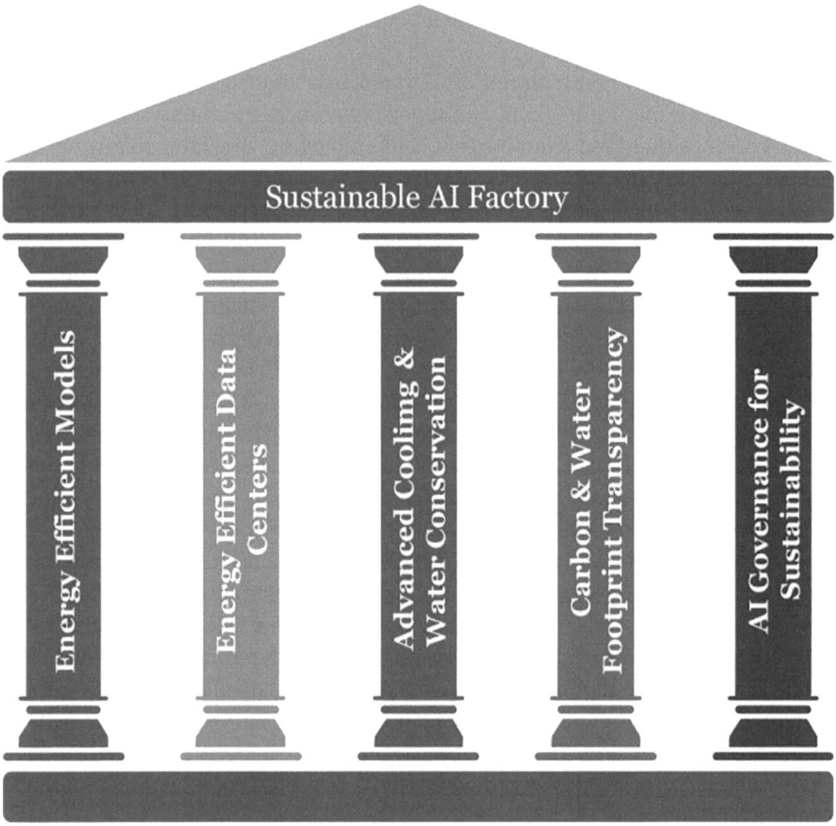

Fig. 7 Pillars of the sustainable AI factory

6 Conclusion and Future Directions

The AIF represents a transformative paradigm in the way organizations conceptualize, develop, and deploy artificial intelligence technologies. Throughout this chapter, we have examined its multifaceted nature, delving into the infrastructure that underpins it, the human-centric principles it champions, the sustainability imperatives it incorporates, and the culture it fosters. These interconnected facets reveal the AIF as both a technological framework and a cultural enabler, designed to seamlessly integrate AI into organizations in ways that amplify value while minimizing risks.

The modular, scalable, and adaptive architecture of the AIF provides a robust foundation for the conceptualization, development, and deployment of AI systems. It is within this structured environment that data is collected, curated, and transformed into actionable insights. The infrastructure not only supports technical workflows, but also promotes collaborative engagement between domain experts, data scientists, and engineers. This ensures that AI systems are not only innovative, but also relevant and effective in addressing real-world challenges.

Beyond its technical aspects, the AIF places significant emphasis on human-centric design. By empowering citizen developers, who possess deep domain knowledge but lack extensive technical expertise, the AIF democratizes access to AI technologies. These individuals can harness AI as a force multiplier, creating solutions that align closely with domain-specific success criteria. Moreover, the AIF fosters cross-functional communication by standardizing terminologies and toolsets. This de-mystification of AI enables project managers, domain experts, and data scientists to collaborate effectively, bridging traditional silos and fostering innovation through shared understanding.

Sustainability emerges as a cornerstone of the AIF's operational philosophy. The environmental dimension underscores the need for energy-efficient computing, carbon footprint reduction, and resource optimization. By adopting green AI principles and leveraging renewable energy sources, the AIF aligns itself with sustainability goals. These can be achieved through scalable architectures, serverless models, and optimized workflows, which ensure cost-effective operations.

The cultural transformation facilitated by the AIF cannot be overstated. Leadership commitment, metrics, and continuous improvement mechanisms foster an organizational culture that embraces AI judiciously. This culture values AI as a trusted colleague, designed to augment human capabilities rather than replace them. By fostering a culture of curiosity, collaboration, and accountability, the AIF ensures that AI technologies are integrated thoughtfully and responsibly.

6.1 *Future Trends*

The potential of the AIF extends far beyond its current capabilities. As technologies evolve, so too will the AIF, incorporating advancements in areas such as explainable AI, federated learning, and neuromorphic computing [27]. These innovations will enhance the transparency, security, and efficiency of AI systems, addressing existing limitations and unlocking new possibilities. The AIF of the future will also deepen its commitment to sustainability by exploring circular economy principles, where resources are reused and recycled to minimize waste.

The success of the AIF ultimately hinges on its ability to balance innovation with responsibility. By embedding human-centric, ethical, and sustainability considerations into every stage of the AI lifecycle, the AIF ensures that technological progress does not come at the expense of societal well-being. This balance requires continuous engagement with stakeholders, robust governance mechanisms, and an unwavering commitment to transparency and accountability (Fig. 8).

In conclusion, the AIF represents a vision for the future of artificial intelligence—one that is innovative, democratized, and sustainable. It is a testament to the potential of AI to drive positive change, provided it is developed and deployed with foresight and care. As organizations embrace the AIF model, they will not only unlock the

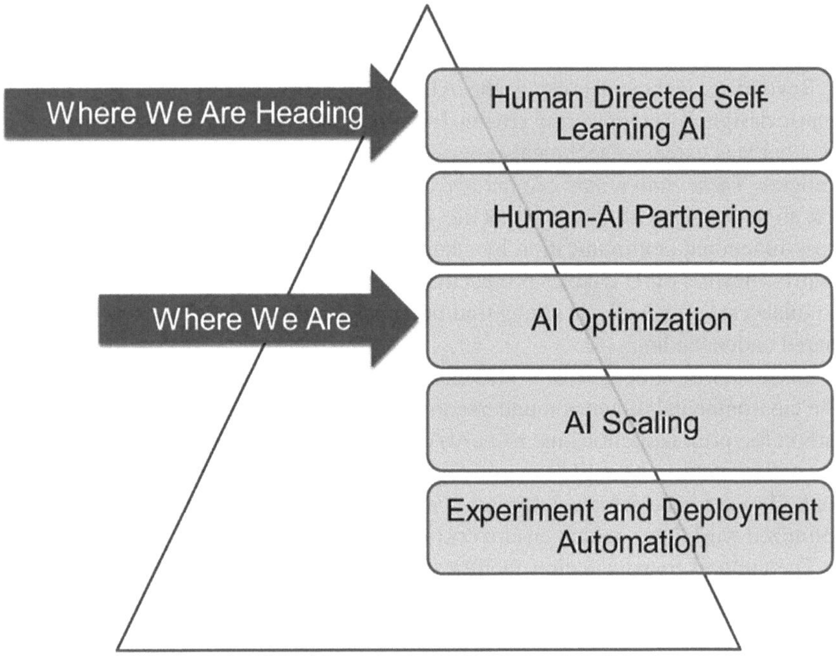

Fig. 8 AI factory vision for the future

full potential of AI but also contribute to a future where technology serves humanity in meaningful and impactful ways. The journey of the AIF is just beginning, and its evolution promises to shape the trajectory of AI for years to come.

References

1. Rashid, A. B. (2024). AI revolutionizing industries worldwide: A comprehensive overview of its diverse application. *Hybrid Advances*, 100277.
2. Karim, R. D. (2023). AI factory: theories, applications and case studies. CRC Press.
3. Mäkinen, S. S. (2021). Who needs MLOps: What data scientists seek to accomplish and how can MLOps help? *EEE/ACM 1st Workshop on AI Engineering-Software Engineering for AI (WAIN)* (pp. 109–112). IEEE.
4. Gal, M. S. (2019). Data Standardization. *NYUL Review 94*, 737.
5. Rodríguez, P. B. (2018). Beyond one-hot encoding: Lower dimensional target embedding. *Image and Vision Computing 75*, 21–31.
6. Abdi, H. &. (2010). Principal component analysis. *Wiley interdisciplinary reviews: computational statistics*, 2(4), 433–459.
7. Olszewski, R. T. (2001). Generalized feature extraction for structural pattern recognition in time-series data. Carnegie-Mellon University.
8. Tabassum, A. &. (2020). A survey on text pre-processing & feature extraction techniques in natural language processing. *International Research Journal of Engineering and Technology*, 7(06), 4864--4867.

9. Salau, A. O. (2019). Feature extraction: a survey of the types, techniques, applications. *IEEE International conference on signal processing and communication*, 158–164.
10. Desmond, M. D. (2020). Semi-automated labeling. *In NeurIPS 2020 Competition and Demonstration Track*, 156–169.
11. Druck, G. S. (2009). Active learning by labeling features. Proceedings of the 2009 conference on Empirical methods in natural language processing, 81–90.
12. Tang, W. &. (2011). Semi-supervised consensus labeling for crowdsourcing. *SIGIR 2011 workshop on crowdsourcing for information retrieval*, 1-6.
13. Yang, L. &. (2020). On hyperparameter optimization of machine learning algorithms: Theory and practice. *Neurocomputing*, 415, 295–316.
14. Naidu, G. Z. (2023). A review of evaluation metrics in machine learning algorithm. *Computer Science On-line Conference*, 15–25.
15. Chen, S. F. (1998). Evaluation metrics for language models.
16. Caton, S. &. (2024). Fairness in machine learning: A survey. *ACM Computing Survey*, 56(7), 1–38.
17. Singh, R. &. (2023). Edge AI: a survey. *Internet of Things and Cyber-Physical Systems*, 3, 71-92.
18. ONNX. (2024). ONNX Specification. *https://github.com/onnx/onnx*.
19. Urblik, L. K. (2024). Containerization in Edge Intelligence: A Review. *Electronics*, 13(7), 1335.
20. Luksa, M. (2017). *Kubernetes in Action*. Simon and Schuster.
21. Leppänen, T. (2021). Data visualization and monitoring with Grafana and Prometheus.
22. Siam, S. I. (2024). Artificial intelligence of things: A survey. *ACM Transactions on Sensor Networks*.
23. Lebens, M. F. (2021). Rise of the citizen developer. *Muma Review*, 5(12), 101–111.
24. Ouchi, W. G. (1985). Organizational culture. *Annual Review of Sociology*, 11(1), 457-–483.
25. Kazim, E. &. (2021). A high-level overview of AI ethics. *Patterns*, 2(9).
26. Van Wynsberghe, A. (2021). Sustainable AI: AI for sustainability and the sustainability of AI. *AI Ethics*, 1(3), 213–218.
27. Schuman, C. D. (2022). Opportunities for neuromorphic computing algorithms and applications. *Nature Computational Science*, 2(1), 10–19.

MIX
Papier aus verantwortungsvollen Quellen
Paper from responsible sources
FSC® C105338

If you have any concerns about our products,
you can contact us on
ProductSafety@springernature.com

In case Publisher is established outside the EU,
the EU authorized representative is:
Springer Nature Customer Service Center GmbH
Europaplatz 3, 69115 Heidelberg, Germany

Printed by Libri Plureos GmbH
in Hamburg, Germany